Graphene

Scrivener Publishing
100 Cummings Center, Suite 541J
Beverly, MA 01915-6106

Advanced Materials Series

The Advanced Materials Series provides recent advancements of the fascinating field of advanced materials science and technology, particularly in the area of structure, synthesis and processing, characterization, advanced-state properties, and applications. The volumes will cover theoretical and experimental approaches of molecular device materials, biomimetic materials, hybrid-type composite materials, functionalized polymers, supramolecular systems, information- and energy-transfer materials, biobased and biodegradable or environmental friendly materials. Each volume will be devoted to one broad subject and the multidisciplinary aspects will be drawn out in full.

Series Editor: Dr. Ashutosh Tiwari
Biosensors and Bioelectronics Centre
Linköping University
SE-581 83 Linköping
Sweden
E-mail: ashutosh.tiwari@liu.se

Publishers at Scrivener
Martin Scrivener(martin@scrivenerpublishing.com)
Phillip Carmical (pcarmical@scrivenerpublishing.com)

Graphene

An Introduction to the Fundamentals and Industrial Applications

Madhuri Sharon and Maheshwar Sharon

Walchand Centre of Research for Nanotechnology and Bionanotechnology, India

Scrivener Publishing

WILEY

Co-published by John Wiley & Sons, Inc. Hoboken, New Jersey, and Scrivener Publishing LLC, Salem, Massachusetts.
Published simultaneously in Canada.

For general information on our other products and services or for technical support, please contact our Customer Care Department within the United States at (800) 762-2974, outside the United States at (317) 572-3993 or fax (317) 572-4002.

Wiley also publishes its books in a variety of electronic formats. Some content that appears in print may not be available in electronic formats. For more information about Wiley products, visit our web site at www.wiley.com.

For more information about Scrivener products please visit www.scrivenerpublishing.com.

Cover design by Russell Richardson

Library of Congress Cataloging-in-Publication Data:

ISBN 978-1-118-84256-0

Printed in the United States of America

10 9 8 7 6 5 4 3 2 1

To Our Grandchildren

ANISH
RACHAEL
ANNIKA
ARYAN

You four mean everything to us; you are our blessings, love and life

Contents

Foreword

Graphene is one of the most incredible materials in that it has just an atom-thinness but has millimeter or even centimeter size area. Before 2004 when the first preparation of single-layer graphene was reported, people had talked about and imagined graphene as ultimately thin "ideal" graphite and also as an "ingredient" of single-wall carbon nanotubes. Since then, partly because of the 2010 graphene Nobel Prize, a number of graphene-related studies has been published worldwide, and it is almost impossible to access and follow every conceivable studies of one's related research area of graphene. The publication of a compact and yet the state-of the-art book on graphene has, therefore, been highly desired and anticipated by researchers.

Professors Madhuri and Maheshwar Sharon have beautifully and successfully realized these requirements by this new monograph. One may be surprised to see how fertile and productive chapters are involved in the book, ranging from basic structures, mechanical/electronic properties of graphene, to various applications of graphene technology and even to graphene in industry and commercialization.

April 2015

Hisanori Shinohara
Department of Chemistry
Nagoya University
Nagoya, Japan

Preface

Science is an ever-continuing quest to understand the intricacies of nature right from atomic scale to vastness of the universe. One of such realm of learning is venturing into materials at a particularly defined size of 1–100 nm—encompassing a science called nano-science and nanotechnology. Graphene is the outcome of research and knowledge based on carbon nanotechnology. Graphene is now at the pinnacle of glorious achievements and has motivated multi-disciplinary research towards developing feasible solutions in various sectors. There have been several advances in the field of graphene-based materials, such as in energy-related applications as fuel cells, super-capacitors and photovoltaic devices. Graphene, by virtue of its unique properties, and graphene composites have also found an important relevance in energy harvesting. Furthermore, applications of graphene in filtering heavy metal ions and other pollutants are also of importance in the current scenario. The recent Nobel Prize–winning research work on graphene has attracted significant attention on account of its exceptional capabilities particularly in the field of electronics.

This book is our humble effort to present the state of the art of graphene research intended for various applications. We have tried to place these developments in scientific, technical, as well as commercial and economic context to assess the likelihood of uptake of these technologies and their relevance to world's pressing needs of energy, miniaturization, communication, transportation and health.

The scope of this book includes scientific and technological details along with present day industrial approach and needs. This book is intended for new entrants and active researchers in the field of graphene science and technology in industry and academia, medical, government officials responsible for research, innovation, entrepreneur and industrialists venturing into applications of graphene, students and interested lay persons. We assume readers have academic training, but no expertise in graphene-technology.

Madhuri Sharon
Maheshwar Sharon
May 2015

1

The History of Graphene

A pencil and a dream can take you anywhere.

Joyce A. Meyers

Prior to excavating the history of graphene, one has to know graphite, which is composed of many layers of graphene stacked together. This stacking makes a three-dimensional structure, the graphite, whereas graphene is a two-dimensional, one-atom-thick material. Evidence of the uses of graphite in Europe has been recorded in pottery decorated with graphite some 6000 years ago. The present concept and clarity about graphite is nearly 500 years old. Graphite ore (Figure 1.1) was found and mined in England in the sixteenth century.

People used graphite to mark their sheep. However, it was believed that this mineral was lead ore and it was called "plumbago". Scheele, in 1779, demonstrated that plumbago is actually carbon, not lead. Because people used it to write marks on their sheep, a German scientist, Verner (1789) named it graphite (a Greek word for "writing"). With the development of the pencil industry, it has been used as a writing material in a pencil (Figure 1.2) since the eighteenth century.

Figure 1.1 Graphite ore. Courtesy: http://en.wikipedia.org/wiki/Graphite.

Figure 1.2 A lead pencil tip made of graphite. Courtesy: http://commons.wikimedia.org/wiki/File:Pencils_hb.jpg.

Because of its layered morphology and weak dispersion forces between adjacent sheets, it was utilized as solid lubricant. Before proceeding further with the history of graphene, it is necessary to define what a graphene is.

The term "graphene" first appeared in 1987 (Mouras *et al.* 1987) to describe single sheets of graphite as one of the constituents. The term "graphite layers" was replaced with "graphene" by the IUPAC commission. According to the recent definition, "graphene is a two-dimensional monolayer of carbon atoms, which is the basic building block of graphitic materials (i.e., fullerene, carbon nano tubes, graphite)". Graphene is a two-dimensional material. It consists of a single layer of carbon atoms arranged in a honeycomb-like structure (Figure 1.3B).

The carbon-carbon bond length in graphene is about 0.142 nanometers (Figure 1.3B). Its layer height was measured to be just 0.33nm (Figure 1.3A). It is the thinnest material known, and yet is also one of the strongest. Graphene is almost completely transparent. Its structure is so dense that even the smallest atom helium cannot pass through it. It conducts electricity as efficiently as copper and outperforms all other materials as a heat conductor.

In 1859 a British chemist, Benjamin Bordie, prepared a highly lamellar structure by thermally reducing graphite oxide by reacting graphite with potassium chlorate and fuming nitric acid, resulting in the formation of a suspension of graphene oxide crystallite. This graphene oxide was later woven into a paper. An early study the properties of this graphene oxide paper was completed by Kohlschutter and Haenni in 1919. Graphene, a molecule arranged in a single atomic plane, is accepted as a two-dimensional crystal. Earlier it was believed it could not be grown, because thermodynamics had been shown to prevent the formation of two-dimensional crystal in free state by Landau (1930).

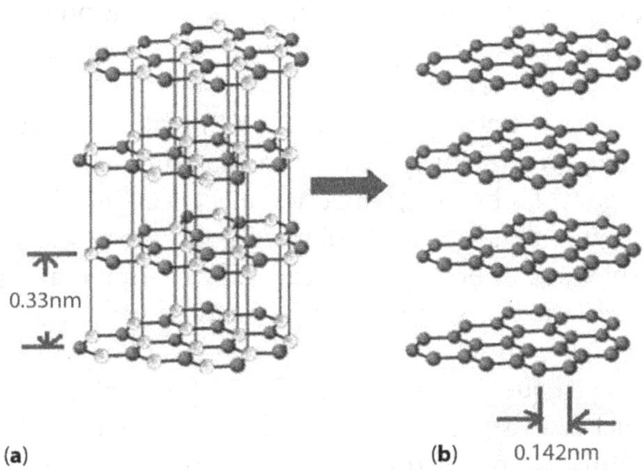

Figure 1.3 Schematic diagram of (a) Graphite and (b) Four layers of graphene from graphite.

Wallace (1947), while trying to study the electronic properties of three-dimensional graphite, came up with the band theory of graphite. According to him,

> The structure of the electronic energy bands and Brillouin zones for graphite is developed using the 'tight binding' approximation. Graphite is found to be a semi-conductor with zero activation energy, i.e., there are no free electrons at zero temperature, but they are created at higher temperatures by excitation to a band contiguous to the highest one which is normally filled. The electrical conductivity is treated with assumptions about the mean free path. It is found to be about 100 times as great parallel to as across crystal planes. A large and anisotropic diamagnetic susceptibility is predicted for the conduction electrons; this is greatest for fields across the layers. The volume optical absorption is accounted for.

The next milestone work regarding graphene was the publication of the first TEM image of a few layers of graphene by Ruess and Vogt (1948).

Ubbelohde and Lewis (1960) isolated a single-atom plane of graphite and reported surprisingly higher basal-plane conductivity of graphite intercalation compounds as compared to that of the original graphite. They pointed out that graphite consists of layers, which are a network of hexagonal rings of carbon atoms.

Hanns-Peter Boehm and his coworkers isolated and identified single graphene sheets by TEM and XRD in 1961. Their work was published in 1962. Boehm also authored the IUPAC (International Union of Pure and Applied Chemistry) report, formally defining the term *graphene* in 1994. It is surprising that many reviews and papers have mentioned that graphene was discovered in 2004. The TEM taken by Boehm *et al.* remained the best observation for over forty years.

These forty years (between 1960 and 2000) exhibited that the research of graphene has grown slowly in multifarious directions, including synthesis. The hope of observing superior electrical properties from thin graphite or graphene layers while obtaining graphene was considered to be a formidable task in both theoretical and experimental aspects. In the graphite intercalation systems, large molecules were inserted between atomic planes, generating isolated graphene layers in a three-dimensional matrix. The subsequent removal of the larger molecules produced a mixture of stacked or scrolled graphene layers without affecting the structure. During this period of research, the cause of the high conductivity of graphite intercalation compounds and the future applications were the main concerns.

There have been attempts to grow graphene using the same approach as the approach generally used for growth of carbon nanotubes, but it allowed the formation of thicker than ≈100 layers graphite films (Krishnan *et al.* 1997).

Hess and Ban (1966) were the first to use a chemical-vapor-deposition (CVD) technique, in which carbon atoms were supplied from a gas phase, to achieve the formation of monolayer graphite or graphene.

However, efforts to epitaxially grow few-layer graphene through the chemical vapor deposition of hydrocarbons on metal substrates (Land *et al.* 1992 and Nagashima *et al.* 1993) and on top of other materials (Oshima and Nagashima 1997) as well as by thermal decomposition of SiC have also been successful.

Epitaxial growth of graphene offers probably the only viable route towards electronic applications and, with so much at stake, rapid progress in this direction is expected. The approach that seems promising but has not been attempted yet is the use of the previously demonstrated epitaxy on catalytic surfaces (Land *et al.* 1992 and Nagashim *et al.* 1993), such as Ni or Pt, followed by the deposition of an insulating support on top of graphene and chemical removal of the primary metallic substrate.

This "epitaxial graphene" consists of a single-atom-thick hexagonal lattice of sp^2 bonded carbon atoms, as in free-standing graphene. However, there is significant charge transfer from the substrate to the epitaxial graphene, and, in some cases, hybridization between the d orbitals of the substrate atoms and π orbitals of graphene, which significantly alters the electronic structure of the epitaxial graphene. The fact that electric current would be carried by effectively massless charge carriers in graphene was pointed out theoretically by Semenoff *et al.* in 1984.

Properties such as the layered morphology and weak dispersion forces between adjacent sheets have made graphite an ideal material for use as a dry lubricant, along with the similarly structured but more expensive compounds hexagonal boronnitride and molybdenum disulfide. High, in-plane electrical (104 Ω^{-1} cm^{-1}) and thermal conductivity (3000 W/mK) enable graphite to be used in electrodes and as heating elements for industrial blast furnaces (Bouchard *et al.* 2001).

The beginning of the twenty-first century saw many important discoveries related to graphene. Enoki *et al.* in 2003 explained the anisotropy of graphite's material properties. Bulk graphite was first intercalated by Dresselhaus and Dresselhaus (2002) so that graphene planes became separated by layers of intervening atoms or molecules. This usually resulted in new three-dimensional materials. However, in certain cases, large molecules could be inserted between atomic planes, providing greater separation, such that the resulting compounds could be considered as isolated graphene layers embedded in a three-dimensional matrix.

Shioyama *et al.* (2001) and Hirata *et al.* (2004) demonstrated that one can often get rid of intercalating molecules in a chemical reaction to obtain a sludge consisting of restacked and scrolled graphene sheets.

Graphene was patented two years before the Nobel Prize Prize–winning work of Andre Geim and Kostya Novoselov (2004) by a company called Nanotek Instruments (US patent number 7071258, entitled "Nano-scaled graphene plates" of 2002, owners, Bor Jang and Wen Huang). This patent includes a sketch of carbon nanotubes unrolling to form graphene sheets and multilayer graphene sheets. It is surprising that this visionary patent is not acknowledged by most graphene researchers today; perhaps because most scientific researchers from the academic world never bother to look into the patent literature, whereas industry leaders tend to follow the scientific literature very closely. We hope that the efforts made in recent years to promote the collaboration between industry and academia would promote the sharing of knowledge.

It is worth mentioning here that Dr. Bor Jang (owner of the first graphene patent) did heaps of work on graphene. He has over forty patents related to graphene production and applications, including the first patent for single layer graphene in 2002 and the first patent on graphene-reinforced metal, glass, carbon and ceramic-matrix composites and single layer graphene-reinforced polymer composites. However, Dr. Jang almost never published scientific papers, for which reason he is almost unknown in academia.

2004 was a golden year for graphene research. There have been a number of efforts to make very thin films of graphite by mechanical exfoliation from 1990 to 2004, but nothing thinner than fifty to 100 layers was produced during these years. In 2004, Andre Geim and Kostya Novoselov at Manchester University, UK, managed to extract single-atom-thick crystallites (graphene) from bulk graphite and transfer them onto thin silicon dioxide on a silicon wafer by a famous Scotch Tape Technique. The idea of using Scotch tape for exfoliating graphene was suggested by Oleg Shklyarevskii, who had been using it to polish the graphite rod of pencils. In this micromechanical Scotch tape exfoliation method, graphene is peeled off from graphite using adhesive tape. Initially multiple-layer graphene gets attached to the sticky tape. Then, folding and peeling the tape several times results in the separation of progressively thinner layers and eventually to a single layer of carbon. To detach the tape, acetone is used. Then one last peeling is performed with unused tape by placing a sample of graphite onto sticky tape. By this method, the best quality of graphene is obtained. However, it is difficult to scale up this method. Other methods, like reduction of exfoliated graphene oxide, are used for scaling up, but the quality of graphene produced is poor. Desorption of Si from SiC or growth on metal both gives good quality graphene and is also scalable.

Though graphene was known earlier, it would not be an exaggeration to write that Geim and Novoselov rediscovered graphene in its new incarnation. Apart from receiving the Nobel Prize in physics in 2010, Geim received several awards for his pioneering research on graphene, including (i) the Mott medal for the "discovery of a new class of materials—free-standing two-dimensional crystals—in particular graphene" in 2007, (ii) the EuroPhysics Prize (together with Novoselov) "for discovering and isolating a single free-standing atomic layer of carbon (graphene) and elucidating its remarkable electronic properties" in 2008, (iii) Körber Prize for "developing the first two-dimensional crystals made of carbon atoms" in 2009 and (iv) in 2010 Geim and Novoselov were granted knighthood.

Geim and Novoselov did the Electric field study of graphene. The silicon beneath the SiO_2 was used as a "back gate" electrode to vary the charge density in the graphene layer over a wide range. Their studies revealed that graphene (monolayer) and even its bilayer have simple electronic spectra; both are zero-gap semiconductors (or zero-overlap semimetals) with one type of electron and one type of hole. For three and more layers, the spectra become increasingly complicated: Several charge carriers appear (Novoselov *et al.* 2004 and Morozov *et al.* 2005), and the conduction and valence bands start notably overlapping (Novoselov *et al.* 2004 and Partoens and Peeters 2006); this led to an explosion of research in synthesis, characterization, properties and research into the potential applications of graphene.

Morozove *et al.* (2005) and Zhang *et al.* (2005) suggested that because the screening length in graphite is only $\approx$5Å (that is, less than two layers in thickness), one must differentiate between the surface and the bulk even for films as thin as five layers. The study allows one to distinguish between single-, double- and few- (3 to <10) layer graphene as three different types of two-dimensional crystals of graphene. Thicker structures are to be considered as thin films of graphite.

In 2005 the Anomalous Quantum Hall Effect was detected, showing the massless nature of charge carriers in graphene by Novoselov *et al.* (2005) and Zhang *et al.* (2005). They used a micromechanical cleavage technique that led them to make the first observation of the Anomalous Quantum Hall Effect in graphene. They also observed evidence of the theoretically predicted π Berry's phase of massless Dirac fermions in graphene. In 1984 DiVincenzo and Mele were the first to point out the massless Dirac equation, and it is known that the magnetic field of an electronic Landau level occurs precisely at the Dirac point. This level is responsible for the anomalous integer quantum hall effect. In 2006, the Anamolous Quantum Hall Effect was observed at room temperature by Novoselov *et al.* in graphene.

Schedin *et al.* (2007) were the first to show the detection of a single molecule adsorption event. They demonstrated that μm-size sensors made from graphene are capable of detecting individual events when a gas molecule attaches to or detaches from graphene's surface. The adsorbed molecules change the local carrier concentration in graphene by one electron, which leads to step-like changes in resistance. The achieved sensitivity is due to the fact that graphene is an exceptionally low-noise material electronically, which makes it a promising candidate not only for chemical detectors but also for other applications where local probes sensitive to external charge, magnetic field or mechanical strain are required.

In 2008 Bolotin *et al.* demonstrated extremely high carrier mobility in suspended graphene. They could achieve mobilities in excess of 200,000 $cm^2 V^{-1}s^{-1}$ at electron densities of $\sim 2 \times 10^{11}$ cm^{-2} by suspending single-layer graphene. Suspension ~150 nm above a Si/SiO_2 gate electrode and electrical contacts to the graphene was achieved by a combination of electron beam lithography and etching. The specimens were cleaned *in situ* by employing current-induced heating, directly resulting in a significant improvement of electrical transport. Concomitant with large mobility enhancement, the widths of the characteristic Dirac peaks were reduced by a factor of 10 compared to traditional, non-suspended devices. This advancement allowed for accessing the intrinsic transport properties of graphene.

2009 was a year of initiation of many commercialization activities for graphene, for example:

- In June—graphene-Info was launched
- In October—IBM researchers used graphene to develop ultra-fast photo detectors
- In November—two industrial companies took up graphene work:
  - Samsung started research in graphene; and
  - Fujitsu entered into the production of graphene transistors at low temperatures
- In December—Angstron Materials awarded $1.5 million to develop nanographene platelets

In 2010 Andre Geim and Kostya Novoselov were awarded the Nobel Prize in physics for their work on graphene.

Figure 1.4 Nobel Laureates Andre Geim and Kostya Novoselov receiving the Nobel Prize for physics for their work on graphene

Graphene has a higher carrier mobility than silicon, but lacks a band gap, which has kept the on-off ratio of graphene transistors dismally low—usually less than 10 compared to hundreds for silicon. In January 2010, IBM researchers created a tuneable electrical band gap (up to 130meV) for their bi-layer graphene field-effect transistors (FET) that could someday rival complementary metal oxide semiconductors. According to IBM, this was one of the last roadblocks to commercialization of graphene-based technology.

By February of the same year, IBM developed a 100-GHz graphene RF Transistor, on 2" wafers, which operates at room temperature, and is more than two times faster than silicon transistors with the same gate length (40GHz). IBM's next aim is to increase the speed of the graphene transistor to 1 THz. The graphene RF transistors were created for the Defence Advanced Research Project Agency under its Carbon Electronics for RF Applications (CERA) program. Transistors were fabricated at the wafer scale using epitaxially grown graphene processing techniques that are compatible with those used to fabricate silicon transistors. Later in 2011, IBM developed a 155 GHz graphene transistor and a 10 GHz graphene based IC.

In September of the same year, UCLA researchers, led by biochemist Xiangfeng Duan, developed a 300 GHz graphene transistor using a nanowire as the self-aligned gate.

In April 2010, the European startup graphenea, with $3.8 million investment, was established to produce graphene, and Xolve, with an investment of $2 million, entered into graphene production.

Samsung, which started research and development on graphene in 2009, managed to fabricate a 30" graphene sheet by roll-to-roll process, and later in 2011, produced a 40" graphene sheet. Samsung is the company that holds the largest amount of graphene patents in the world.

Wei Han *et al.* (2010) achieved tunneling spin injection from Co into single-layer graphene (SLG) using TiO_2 seeded MgO barriers. A non-local magneto-resistance (ΔR_{NL}) of 130 Ω at room temperature was observed, which is the largest value observed in any material. Investigating ΔR_{NL} vs. SLG conductivity from the transparent to the tunneling contact regimes demonstrates the contrasting behaviors predicted by the drift-diffusion theory of spin transport. Furthermore, tunnel barriers reduce the contact-induced spin relaxation and are therefore important for future investigations of spin relaxation in graphene.

In 2011 Vorbeck Materials entered into making graphene-based ink to develop a Siren alarm (security) tag, the world's first graphene-based ink product. By 2012 they started shipping it.

Realizing the importance and potential for application of graphene, many government agencies started investing in graphene research in 2011.

- Sweden granted $6 million for graphene research.
- The UK government decided to invest £50 million in graphene commercial opportunities. In 2012, the UK gave £21.5 million more for graphene research. In 2013 the UK government funded £5 million pounds for graphene membrane research at the University of Manchester. In the beginning of 2013 (January), the University of Cambridge established a new £12 million graphene center, and signed a co-development agreement with Plastic Logic in August 2013. In the same year, ERDF awarded £23 million to Manchester University's NGI (National Graphene Insist). Moreover, UK launched a $1.5 million project to develop graphene-filled epoxy resin.
- Korea allocated $40 million towards graphene commercialization efforts in 2013.
- MIT (USA) opened a new center for graphene devices and systems (MIT-CG).
- The EU decided to grant €1 billion for graphene research initiatives over 10 years.

2011 saw another applicable output of graphene. Dickerson's group created a film of graphene oxide that either causes water to bead up and "run off" or alternatively be spread out in a thin layer. As graphene sheets are transparent, this film can be used in (a) car windshields, where water will shed so quickly that wipers will not be needed, (b) to make ships glide through water very efficiently, (c) to make water-repellent clothes, or (d) self-cleaning glasses, which were developed by Dickerson's group.

Researchers at the Korea Advanced Institute of Science and Technology (KAIST) used graphene to create bendable batteries. The researchers developed a graphene-based hybrid electrode and produced a flexible lithium rechargeable battery. The cathode material (V_2O_5) was grown on a graphene sheet using pulsed laser reposition and the anode was lithium-coated graphene. This battery has promising performance compared to non-flexible batteries—higher energy density, power density and a better cycle life. They also believe that this technology can be used not only in batteries but also in solar cells, OLED displays and catalysis.

2012 and 2013 saw much commercial growth and developments in graphene technology, for example:

- Bluestone Global Tech started producing 24" by 300" graphene films on copper.
- HEAD developed a graphene racket.
- Vorbeck and PNNL started marketing graphene-based lithium-ion batteries; and
- Cabot launched a graphene-based additive for high density lithium-ion batteries.
- UCLA entered into the commercialization of laser-scribed graphene supercapacitors.
- Grafoid started to mass produce affordable high-quality graphene materials called *MesoGraf* with an investment of $3.5 million.
- Durham graphene Science raised £1.2 million to commercialize graphene mass production.
- IBM developed a graphene-based Terahertz frequency photonic filter and polarizer.
- Sony could produce a 100-meter long graphene sheet with a new R2R method.
- The National University of Singapore invested $11 million in a graphene production facility.
- Haydale announced the availability of HDPl as graphene-based inks.

- Lockheed Martin developed a new graphene-based water desalination technology, which they hope to commercialize by 2014–2015.
- China launched production of 15" graphene transparent film.

2012 and 2013 witnessed many milestones in research as well.

A breakthrough in the photovoltaic cell was announced by MIT researchers in 2012. They developed a flexible graphene-based solar panel (flexible, light solar cells): Researchers developed a new approach using graphene sheets coated with nanowires.

This new solar (photovoltaic) cell is made from several graphene sheets coated with nanowires. Here, graphene is used as a replacement for ITO in the solar panels. The new electrode material is cheaper and provides several advantages over ITO: Flexibility, low weight, mechanical strength and chemical robustness. A series of polymer coatings to modify the graphene properties, allowing them to bond a layer of zinc oxide nanowires to it, and then an overlay of a material that responds to light waves, i.e., either lead-sulphide quantum dots or a type of polymer called P3HT, is used. Despite these modifications, graphene's innate properties remain intact. A solution-based process was used to deposit the zinc oxide nanowires on the graphene electrodes, which can be done entirely at temperatures below 175°C. This process is highly scalable. The graphene itself is synthesized through CVD and then coated with the polymer layers.

Improving solar cells demands (a) increased efficiency of their energy conversion, (b) lowering the cost of manufacturing and (c) now Jeffrey Grossman's team at MIT are opening another avenue for improvement, aiming to produce the thinnest and most lightweight solar panels possible. They have come up (2013) with a graphene/molybdenum solar cells to achieve "ultimate power conversion," which is made from a stack of two one-molecule-thick materials: Graphene and molybdenum disulfide. They claimed that these two sheets together are thousands of times thinner than conventional silicon solar cells.

Singh *et al.* (2011) of Monash University, Australia, published a report suggesting that graphene coating can make copper almost 100 times more resistant to corrosion. Graphene coating on copper (Cu) increased the resistance of the metal to electrochemical degradation by one and a half orders of magnitude. Detailed electrochemical characterization in aggressive chloride environment study showed that the impedance of Cu dramatically increases and the anodic and cathodic current densities of the coated Cu becomes nearly one to two orders of magnitude smaller when coated with graphene. The observations are counterintuitive, as graphite in contact with metal, increases metallic corrosion. The results are expected to bring paradigm changes in the development of anti-corrosion coatings using conformal, ultrathin graphene film.

Another graphene miracle surfaced in 2012 when a team led by Novoselov showed that graphene can repair itself by fixing holes in its structure if it is exposed to loose carbon atoms.

Seon Jeong Kim's group in 2012 (Min Kyoon Shin *et al.* 2012) created the world's toughest fibers by mixing a polymer sheet with reduced graphene oxide and carbon nano tubes during spinning.

Researchers from the Georgia Institute of Technology managed to create a substantial bandgap in graphene nanoribbons by coating bilayer graphene on silicon carbide nanometer-scale steps. This could lead the way towards graphene-based electronics.

Graphenea has produced up to 50% stronger ceramic by adding graphene to ceramic alumina by a simple, fast and scalable method. Moreover, it makes the alumina a hundred million times electronic more conductive to electricity.

Recently, graphene's negative resistance–enabled bandgap-less transistor has been designed by Guanxiong Liu and his group from the University of California, Riverside, who developed a graphene-based transistor based on negative resistance rather than trying to open up a bandgap. Negative resistance is the counterintuitive phenomenon in which a current entering a material causes the voltage across it to drop. It was previously demonstrated that graphene demonstrates negative resistance in certain circumstances.

IBM researchers are working on various applications of graphene. They have developed a graphene-based infrared detector, driven by intrinsic plasmons, which is much more photo-responsive compared to non-plasmonic graphene detectors. CVD technique was used to grow graphene on copper foil. The copper was etched away and the graphene sheet was transferred to a silicon/silicon-oxide chip. The researchers patterned graphene ribbons (widths of 80 to 200 nm). Graphene is less useful than a semiconductor to detect light in the visible range, but in the infrared range (and also in the terahertz range) graphene is very good, as its high mobility and zero-gap nature gives it fast optoelectronic response and detection. In April 2013, IBM discovered that plasmons lose their energy very slowly in graphene, which is good for photonics and quantum optics applications.

Johan Liu, who heads the international (collaborators are Hong Kong University of Science and Technology, Shanghai University in China and Swedish company SHT Smart High Tech AB) research project, has proposed a new energy-saving application of graphene. According to the researchers, graphene can reduce the working temperature in hotspots inside a processor by up to 25%, which can significantly extend the working life of computers and other electronics and reduce the amount of energy used on cooling.

Modern electronic systems generate a great deal of heat, above all due to the constantly increasing demand for more and more functionality. It is important to be able to remove the heat generated in an efficient way to maintain the long life of the system. One rule of thumb is that a 10-degree celsius increase in working temperature halves the working life of an electronics system.

Alex Zettl *et al.* of University of California, Berkeley, have developed a new graphene-based earphone-sized speaker that can actually outperform the best earphones. They say that even with almost no specialized acoustic design, it performs comparably to high-quality commercial headsets. This graphene loudspeaker has a frequency response across the entire audio frequency range (20 Hz–20 kHz). Loudspeakers work by vibrating a thin diaphragm. These vibrations then create pressure waves in surrounding air that produce different sounds depending on their frequency. The human ear can detect frequencies of between 20 Hz (very low pitch) and 20 KHz (very high pitch), and the quality of a loudspeaker depends on how flat its frequency response is—that is, the consistency of the sound it produces over the entire 20 Hz–20 KHz range. The Berkeley team says that its CVD technique for fabricating the speaker is very straightforward and could easily be scaled up to produce even larger-area diaphragms and thus bigger speakers. According to Zettle, "The configuration we describe could also serve as a microphone."

Recently, a collaborative effort by team leader Grigorieva of the University of Manchester with researchers from Finland, Singapore and the Netherlands, has resulted in the discovery of another surprising property of graphene. Graphene was made to be magnetic and its magnetism could be switched on and off at the press of a button, opening a new avenue towards electronics with very low energy consumption. This is the first time magnetism itself has been toggled, rather than the magnetization direction being reversed. Dr. Grigorieva and her team have shown that the magnetic clouds can be controllably dissipated and then condensed back. She explains: "This breakthrough allows us to work towards transistor-like devices in which information is written down by switching graphene between its magnetic and non-magnetic states. These states can be read out either in the conventional manner by pushing an electric current through or, even better, by using a spin flow. Such transistors have been a holy grail of spintronics."

Mark Hersam, professor of materials science and engineering at Northwestern's McCormick School of Engineering and Applied Science, has developed a highly conductive graphene ink for inkjet-printer. To make the graphene ink, graphene was exfoliated from graphite flakes in dimethylformamide (DMF), and then DMF was exchanged by terpineol through distillation (there is a large difference between DMF's and terpineol's boiling points). Terpineol used is of a much lower volume than DMF and so the graphene is significantly concentrated. Terpineol is also non-toxic and features high-viscosity (Secor *et al.* 2013).

The breakthrough in this research is the development of polymer stabilization that achieves a good ink formulation, which makes it stable. The stabilization was achieved by adding a small amount of polymer (ethyl cellulose) before distillation. After printing, the stabilizing polymers can be effectively removed through a simple annealing process.

Apart from electronics applications, graphene is also being used in health care and cosmetics. Recently, scientists from Ulsan National Institute of Science and Technology (UNIST), Korea, have embedded a LED inside a regular soft contact lens, using transparent and conductive electrodes made from graphene and silver nanowires. This is the first time an electronic device was embedded inside a contact lens using flexible and transparent materials.

In this chapter, efforts were made to give a bird's eye view on the different aspects of graphene and to demonstrate how this material has become more interesting than graphite. Political aspects of graphene are also touched upon. Some specific method of its production is highlighted. In the subsequent chapters each aspect of the points discussed here shall be elaborated in very great detail.

2

Structure and Properties of Graphene

All matter originates and exists only by virtue of a force... We must assume behind this force the existence of a conscious and intelligent mind. This mind is matrix of all matters.

Max Planck

2.1 The Structure of Graphene

Graphene is a single layer of carbon atoms forming a perfect honeycomb lattice; it is a two-dimensional structure. The honeycomb network of graphene is composed of sp^2 hybridized carbon atoms. Multi-layers of stacked graphene forms the much-used graphite. Because graphene is a pure carbon material, before we understand graphite or graphene, it is important to understand the basic physics of carbon.

2.1.1 Carbon

Carbon (C) is the sixth element in the periodic table, with six electrons surrounding its nucleus. It is the sixth most abundant element found on Earth. The electronic structure of carbon can be represented as $1s^2\ 2s^2\ 2p_x^{\ 1}2p_y^{\ 1}\ 2p_z^{\ 0}$. The energy level distribution of electrons in carbon is shown in Figure 2.1. Some wonder why we show only $p_z^{\ 0}$ with no electron. In fact, energy levels of p_x p_y and p_z are equivalent, as shown in Figure 2.1. Hence, for our convenience and to look symmetrical we keep p_z with no electron.

Out of its six electrons, four electrons in its valence shell play a significant role in forming three types of hybridizations, namely *sp*, *sp*^{2} and *sp*^{3}. This is very uncommon with other elements.

Figure 2.2 shows the pictorial view of graphite, along with different hybridization seen in carbonaceous materials. *sp*^{3} and *sp*^{2} bonding carbon atoms are the main building block of all stable carbonaceous structure available to date.

Quantum physicists and Nobel Prize in physics winners of 1932 and 1933, Schrödinger, Heisenberg and Dirac, described how the six electrons behave under the influence of the carbon nucleus.

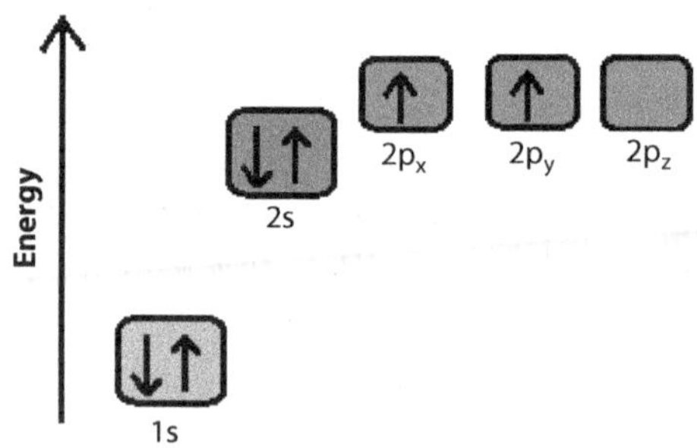

Figure 2.1 Energy levels of electrons occupying 1s 2s 3p orbitals in carbon atoms.

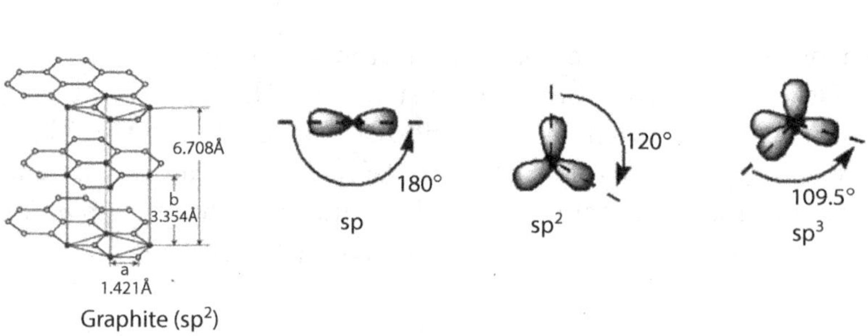

Figure 2.2 Various hybridizations of carbon.

Erwin Schrödinger

Werner Heisenberg

Paul Adrian Maurice Dirac

Carbon has the unique capacity of forming diverse organic compounds. It also possesses an unusual polymer-forming ability. It can bond with many elements in many different ways. No wonder it is the basic element of all known life forms. In its atomic form, carbon is very short-lived; therefore, it is stabilized in various multi-atomic structures with different molecular configurations or allotropes, such as amorphous carbon, graphite, diamond and newly discovered fullerenes in 1985, which include buckyballs, carbon nanotubes, carbon nanobuds, nanofibers, and so on.

2.1.2 Graphite

Carbon's ability to create new materials and even life is due to its unique ability to form long chains, having a central core of carbon atoms. In graphite, carbon forms σ bonds (i.e., covalent bonds) between two adjacent carbon atoms having a carbon-carbon bond length of approximately 0.1421 nm (Figure 2.2).

By sharing three sp^2 electrons with three neighboring carbon atoms, a layer of honeycomb network of planar structure is formed. Graphite is composed of many such stacked planar structures with two different types of stacking arrangements, namely hexagonal and rhomboidal (Figure 2.3), and the fourth electron of carbon is delocalized over the whole plane of graphite. This sp^2 carbon has very strong in-plane bonding (covalent bond) but weak, out-of-plane bonding (van der Waals forces). Due to this weak inter-planar bond spaced at a distance of 0.335nm, graphite sheets can slide onto each other, thus forming graphene.

John D. Bernal (1926) was the first to propose that graphite has a lamellae structure of two types: Hexagonal and rhombohedral carbon rings (Figure 2.3).

In a hexagonal structure, each carbon atom is symmetrically related to its three nearest neighboring carbon atoms, which are in the same plane. As mentioned above, the in-plane carbon-carbon distance is 0.1421 nm, an in-plane lattice constant a_0 of 0.2456 nm, and c-axis lattice constant c_0 of 0.6708 nm and inter-planar distance ao is 0.3354 nm. The planar ring is stabilized by resonance and delocalization of the electrons in the plane. This structure is formed when the planar sheets stack in the order ABABAB, such that all atoms lamellae lie above/below each other and similarly with lamellae B (Figure 2.3).

In a rhombohedral structure, the lamellae are arranged in the sequence ABCABC (Figure 2.3). In both cases the stacking distance between sheets is 0.3354 nm. The hexagonal form is more stable than the rhombohedral.

A schematic of the in-plane σ bonds and the π orbitals perpendicular to the plane of the graphene sheets in graphite is shown in Figure 2.4.

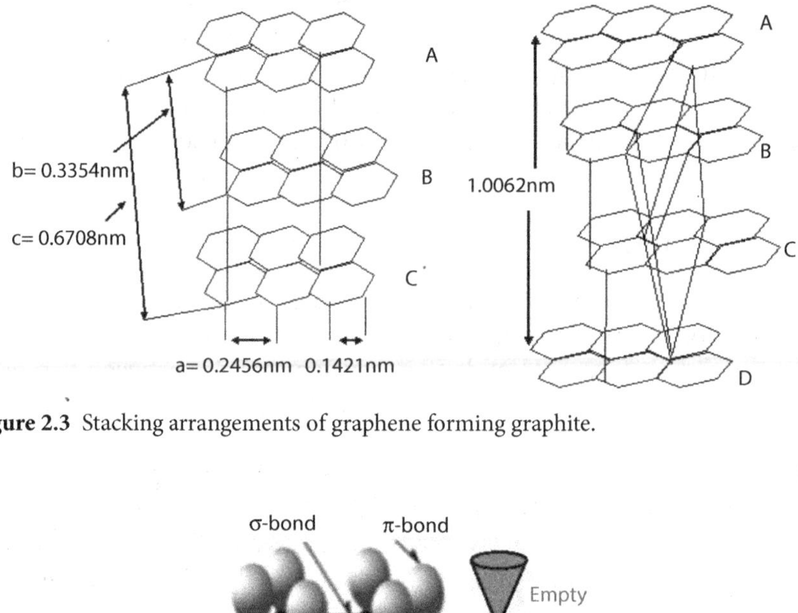

Figure 2.3 Stacking arrangements of graphene forming graphite.

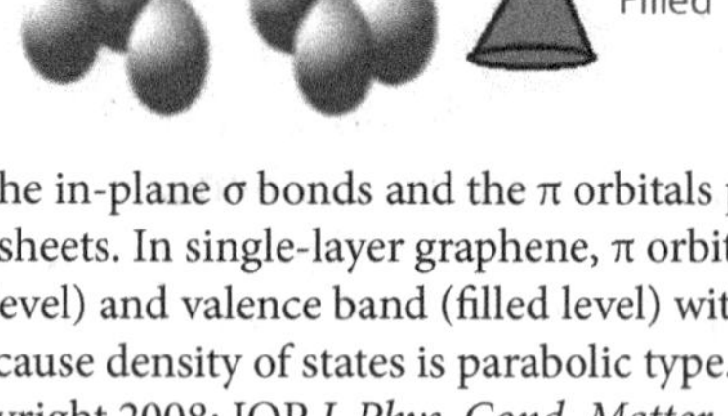

Figure 2.4 Schematic of the in-plane σ bonds and the π orbitals perpendicular to the plane of the graphene sheets. In single-layer graphene, π orbitals overlap, forming conduction band (empty level) and valence band (filled level) with zero band gap. A conical shape is shown because density of states is parabolic type. Copyright 2008: IOP; Courtesy: Hass *et al.*, Copyright 2008: IOP *J. Phys. Cond. Matter*, 20, 323202 (2008).

2.1.3 Graphene

Boehm (1962) described and coined the word graphene as an allotrope of carbon of single planar sheet of sp^2-bonded carbon atoms, densely packed in a honeycomb. The structural relationship between graphite and various newly developed forms are correlated, as can be seen in Figure 2.5. Rather, the honeycomb structure is the basic building block of other important allotropes of carbon, for example:

i. Stacked honeycomb structure forms three-dimensional graphite,
ii. When arranged in two-dimensional structure, it forms graphene,
iii. Rolled honeycomb structure forms one-dimensional carbon nanotubes, and
iv. Wrapped honeycomb structure forms zero-dimensional fullerenes.

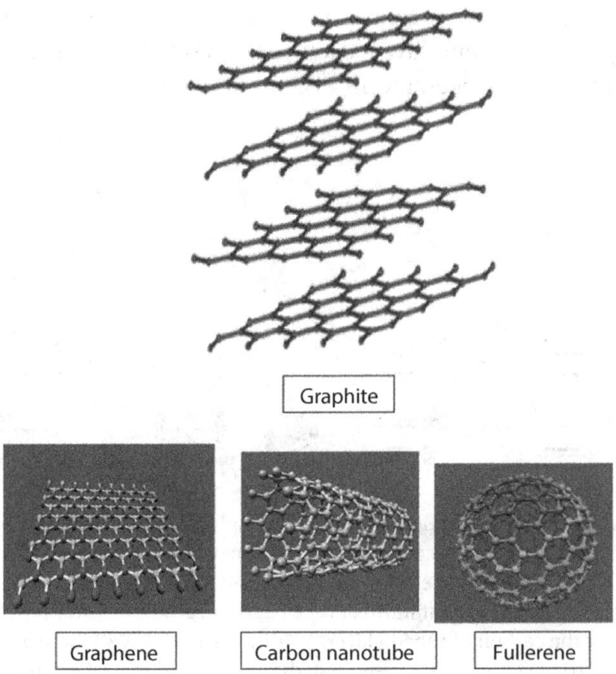

Figure 2.5 Various forms of carbon originating from the honeycomb lattice arrangement of graphite.

Graphene sheets contain in-plane σ-bonds and out-of-plane π-bonds (Figure 2.6A). The π-bonds contribute to the electron conduction of graphene and provide a weak interaction between graphene layers. The covalent σ-bonds form a rigid backbone of the hexagonal structure and the c-axis plane, i.e., π-bonds control communication between different graphene layers. It shows three σ-bonds/atoms in one plane and π-orbitals perpendicular to the plane σ- bonds/atom.

2.1.3.1 *Bilayer Graphene*

Bilayer graphene is two layers of graphene. It can be formed either in twisted configurations (i.e., two layers are rotated relative to each other) or graphitic Bernal stacked configurations (i.e., when half atoms in one layer lie atop half the atoms in the other). Bilayer- and few-layer-graphene are classified as pseudo-2D sp^2 hybridized carbon structures. This graphene exhibits properties that are different from monolayer graphene and graphite.

In bilayer graphene, carbon atoms are stacked in unique ways, hexagonal or AA stacking (Figure 2.6B), or AB stacking (Figure 2.6C). It is worth noting that the end of each hexagonal structure of graphene layer (mono or bilayer) will contain hydrogen atoms to satisfy its valence of four. This means that in a graphene layer, each hexagonal structure at its interface will contain three hydrogen atoms.

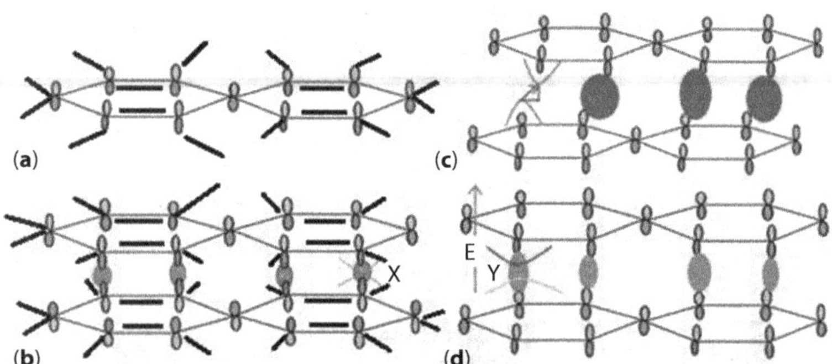

Figure 2.6 (a) Single-layer graphene; (b) Bilayer graphene with AA-stacking; out of plane the π-bond of top layer of graphene overlaps with the π-bond of the lower layer, forming a band separated by a zero distance –X; (c) Bilayer graphene of AB stacking, having similar band formation, zero band gap of indirect type because carbon of lower layer does not lie exactly below the carbon of the top layer, as is the case with AA type structure; and (d) when an electric field in c- axis is applied, there is a slight increase in the band gap -Y by about 250meV; a black line shows the σ-bond for another carbon. It also shows the presence of two double bonds per hexagonal structure.

In bilayer graphene, band gap can be created by applying an external electrical field. Its magnitude can be controlled by the magnitude of the electric field. It is possible to change the band gap of bilayer graphene from zero to infrared energies. Moreover, band gap can be precisely tuned from zero to infrared energies, allowing great flexibility in the design and optimization of semiconductor devices. In Figure 2.6 B and D, this effect is shown.

In monolayer graphene there is no place for the p_z orbital (i.e., π bond) to interact with the carbon atom underneath (because it is absent), whereas in bilayer graphene (Figure 2.6B) such possibility exists. Due to this possibility, it forms a band gap with zero energy. If an electric field is applied in c-axis, there is a possibility that it can shift the band gap from zero to a maximum of 250meV. This shift in band gap can be manoeuvred with the magnitude of the electric field (Figure 2.6D). This condition is favored when the bilayer has a stacking like AA type. If the stacking is of AB type, the p_z of both carbon atoms are not exactly below each other. Hence, such an interaction would cause the formation of indirect band gap. Its magnitude may also be able to change from zero to 250meV. There is a need to establish this behavior with AB type graphene. In addition, it is also necessary to establish whether the shift in the band gap is due to change in the distance 0.3354 nm, or is due to shift in the magnitude of wavelength of π-electrons.

2.1.4 Graphane

Hydrogenation of graphene has led to the discovery of two new forms of graphene: Graphane and graphone. Graphane is a two-dimensional polymer of carbon and hydrogen, e.g., $(CH)_n$ where n is a large number. As mentioned previously, the graphene layer is also hydrogenated, but only those carbons that are at the interface with a non-carbon environment, which may be air. Full hydrogenation from both sides of the graphene sheet results in formation of graphane, whereas partial hydrogenation leads to hydrogenated graphene. Similarly, both-side fluorination of graphene (or chemical and mechanical exfoliation of graphite fluoride) leads to fluorographene (graphene fluoride), while partial fluorination (generally halogenations) provides fluorinated (halogenated) graphene. This means that carbon bonds are in sp^3 configuration, as opposed to graphene's sp^2 bond configuration (Figure 2.7). Thus, graphane can be considered as a two-dimensional analog of cubic diamond.

The first theoretical description of graphane was reported in 2003. Its preparation was reported in 2009. The term graphane applies when all

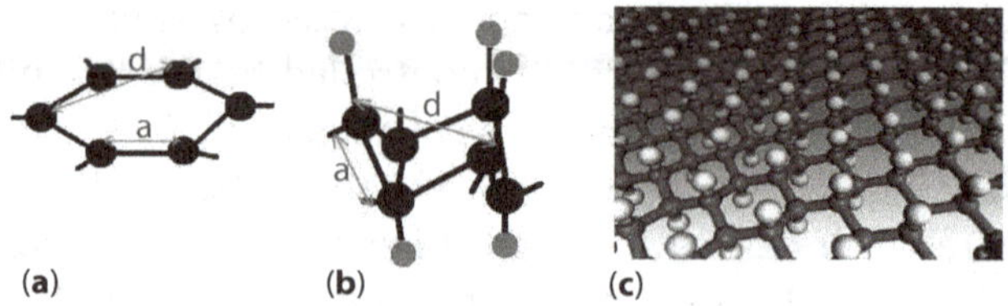

Figure 2.7 Schematics of (a) The hexagonal structure of graphene (b) Changes in the hexagonal structure of graphene after hydrogenation showing graphane having chair configuration (c) Sheet of graphane (the carbon atoms are shown in gray and the hydrogen atoms in white); showing diamond-like distorted hexagonal network with carbon in sp_3 hybridization. Courtesy (C) Copyright @ 2007American Physical Society Jorge *et al.*, *Physical Review B*,75(15), 153401, 2007.

graphene's bonds are saturated with hydrogen atoms. Graphane is synthesized by electrolytic hydrogenation of graphene, few-layer graphene, or mechanical exfoliation of hydrogenated top layers of pure pyrolytic graphite.

2.1.5 Graphone

Because graphone is half-hydrogenated graphene, it can be described as a carbon sheet with a twist. Bilayer graphone, which is nonmagnetic and metallic due to delocalized orbital on the carbon chains, has been expected to have potential application for "spin transport electronics" by controlled hydrogenation. Spintronics is also known as magneto-electronics, where the property of spin of electron, its associated magnetic moment and charge are used for developing solid-state devices, which can store and transfer information in digital form.

In the triangular graphone (Figure 2.8A), all electrons are localized in the hydrogen atoms. No electrons are found in the space between any two other hydrogen atoms. Thus between any two hydrogen atoms there is a very weak interaction, making triangular graphone metastable.

In the rectangular graphone (Figure 2.8 B) electron density between any two hydrogen atoms is large, making a strong interaction between hydrogen atoms. The electronic structure calculation indicates that rectangular graphone is anti-ferromagnetic, possessing an indirect band gap of 2.5eV in comparison to triangular graphone, which is ferromagnetic with a very small band gap of nearly zero.

Bond lengths of C-C and C-H in triangular graphone as shown in Figure 2.9 A and B are 1.495 and 1.157 Å respectively. The distance between

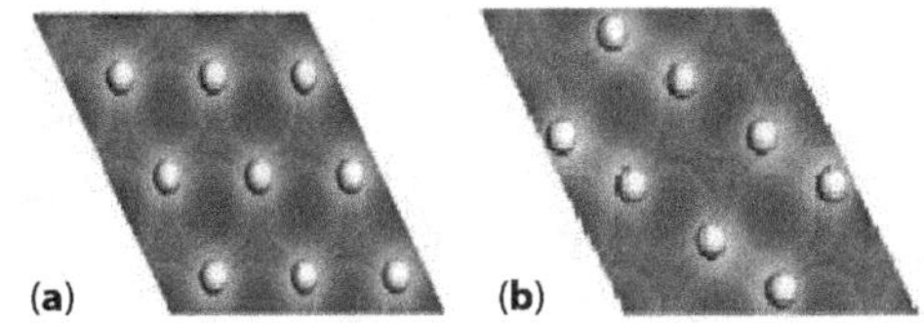

Figure 2.8 Different types of graphone sheets having ferromagnetic properties (a) Triangular graphone and (b) Rectangular graphone. Courtesy: L. Feng, W. X. Zhang, *AIP Advances*, 2, 042138, 2012; doi: 10.1063/1.4766937.

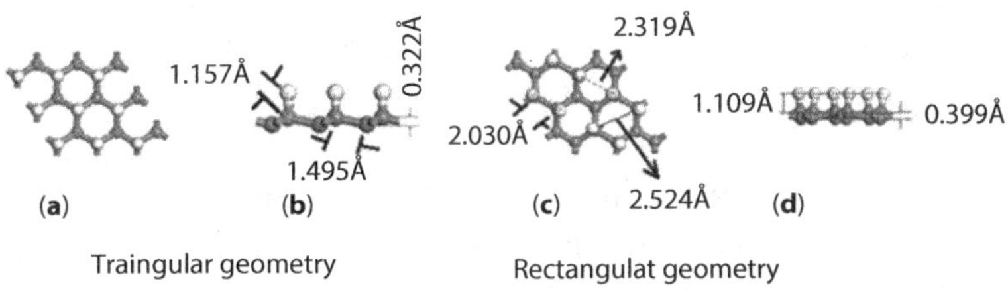

Figure 2.9 Geometry of (a) Triangular graphone (b) Side view of the same showing C-C and C-H distances (c) Rectangular graphone top view shows the unit cell and the distances between H atoms (d) Side view shows distance between C and H planes. (Courtesy: L. Feng, W. X. Zhang, *AIP Advances*, 2, 042138, 2012; doi: 10.1063/1.4766937)

two C planes is 0.322 Å. On the other hand, bond length of C-H in rectangular graphone is 1.109 Å, which is smaller than the C-H bond length in triangular graphone. Moreover, the bond of C-H is not perpendicular to C planes. All the H atoms are in the same plane. The distance between the H atoms are 2.030, 2.524 and 2.319 Å, respectively. The distance between two C planes is 0.399 Å. This distance is a little larger than that of triangular graphone.

2.2 Disorder in Graphene Structure

Understanding graphene structure will remain incomplete, unless we look at the existing and enforced disorders or defects in them. These disorders or the presence of lattice imperfections and oxygen functionalization of graphene declines its quality, but it has contributed immensely in enhancing and understanding the unique and novel properties of graphene. However, clean and highly ordered graphene exhibits high carrier mobility at room temperature (Novoselov *et al.* 2004; Geim and Novoselov 2007) and also

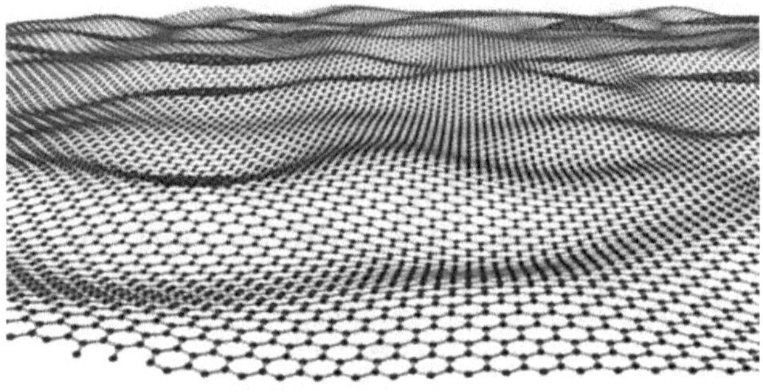

Figure 2.10 Graphene sheet with ripple topography. Courtesy: Copyright © 2008, American Chemical Society Jannik C *et al.*, *Nano letters*, Nov 1 2008.

thermal conductivity (Balandin *et al.* 2008). These properties are very sensitive to defects and disorders, and they become electrically and thermally insulating (Teweldebrhan 2009). Impurities like ad-atoms or defects create magnetic moments in graphene. Types of defects can be categorized as follows:

2.2.1 Ripples

Ripples are an intrinsic feature of graphene sheets and are expected to strongly influence electronic properties by inducing effective magnetic fields and changing local potential number of flexural modes per unit of area at a certain temperature T, L_T, is the thermal wave length of flexural modes.

$$N_{ph} \approx \frac{2\pi}{L_T^2} \ln\left(\frac{L}{L_T}\right), \quad L_T = \frac{2\pi}{\sqrt{k_B T}}\left(\frac{\kappa}{\sigma}\right)^{1/4}$$

At T = 300K, L ~1Å; Indicating that free-floating graphene should always crumple at room temperature due to thermal fluctuations associated with flexural phonons (Neto *et al.* 2009).

2.2.2 Topological Defects

The presence of pentagons, heptagons, and their combination such as Stone-Wales defects in the honeycomb lattice of graphene are called topological lattice defects (Figure 2.11).

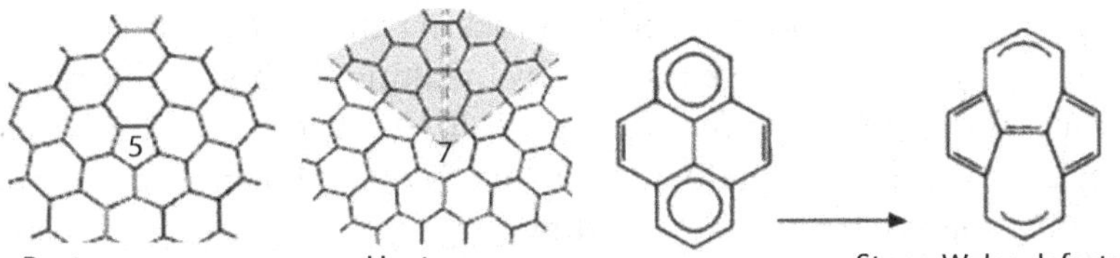

Figure 2.11 Topological defects in graphene. Courtesy:. Copyright © 2010, American Physical Society, Oleg V. Yazyev and Steven G. Louie, *Physical Review B*, 81, 195420 2010.

2.2.3 Ad-Atom (OR "ADSORBED ATOM")

According to http://en.wikipedia.org/wiki/Adatom

> An Adatom is an atom that lies on a crystal surface, and can be thought of as the opposite of a surface vacancy. This term is used in surface chemistry, when describing single atoms lying on surfaces and surface roughness.

Metal adatoms have been found to cause destruction of graphene. Faye (2012) has studied the adsorption of metal adatoms K and Cu on graphene, using density functional theory (DFT) and generalized gradient approximation (GGA). These studies have shown that the interaction between alkali-metal adatoms and graphene has a large ionic character. Moreover, transition metal adatoms or impurities, such as iron, nickel and especially cobalt dramatically reduce the vacancy formation energies, whereas gold impurities have almost no effect on characteristics of the vacancies (Boukhvalov and Katsnelson 2009). Obviously graphene needs protection from transition metals so that destruction is not caused. The addition of adatoms, although it causes defects, also induces magnetic properties.

2.2.4 Cracks or Fractures in Graphene

Cracks in graphene is a significant consideration when electronic devices using graphene are produced. To investigate cracks and crack-paths in single-layer graphene, molecular dynamics (MD) simulation is used. It has been found that crack-path strongly depends on the orientation of the initial crack (Omeltchenko *et al.* 1997), which depends on the use of the cut-off function, in the potential (Khare *et al.* 2007). Another factor that governs the fracture of single-layer graphene is the competition between bond breaking and bond rotation at a crack tip. However, symmetric cleavage fracture can be induced by breaking the first bond at the crack tip, and asymmetric cleavage fracture by breaking the bond adjacent to the first bond at the crack tip (Terdalkar *et al.* 2010). Another method used for studying the crack propagation is by using a coupled quantum/continuum mechanics approach. It helps in finding the crack propagation in armchair and zigzag graphene sheets. This method exhibited that initially the crack is a perpendicular zigzag, and armchair edges, which eventually grow as irregular in armchair sheets but continue as self-similar crack growth in zigzag sheets.

Because cracks in graphene are a factor that alter or affect the uses where the high mechanical strength of graphene is a consideration. The velocity and instability of crack motion in the hexagonal lattice of graphene under pure opening loads needed investigation, which was done by atomistic molecular dynamics simulations. It was found that the brittle crack in zigzag direction can propagate supersonically at even 8.82 km/s under uniform normal loading of edge displacements. Crack moving in straight line at low speed produces atomically smooth edges, while kinking occurs beyond a critical speed around 8.20 km/s equivalent to 65% of Rayleigh wave speed in graphene.

2.3 Properties of Graphene

Humankind has always been intrigued by the behavior of various mechanisms at the sub-atomic level. A deep insight and understanding of many newly discovered materials at a much smaller level, mainly due to the invention of instruments to peep into the world of atoms, molecules and the change in the properties of materials when they combine, in many ways became clearer. This congregation of molecular arrangement has given birth to one of the materials with the most potential of the century, graphene. The cause for the specific properties of graphene stem from the weird world of their chemical configuration, which can be exploited for a plethora of commercial and fundamental applications. Graphene has exhibited very complex compositions, depending on what interactions it has with other chemicals, e.g., hydrogen. Various types of graphene, namely unilayer graphene, bilayer graphene, graphane and graphone, have novel properties, which differ remarkably from the graphite. Monolayer graphene was first prepared from graphite and its properties were studied by Novoselov *et al.*

In the following paragraphs, wide varieties of properties displayed by graphene are discussed.

Table 2.1 Mechanical properties of graphene

Graphene	Mechanical Property	Assessed By	References
Graphene	Strain~ 1.3% in tension Strain~ 0.7% in compression	Raman Spectroscopy	Tsoukleri *et al.* 2009
Monolayer Graphene	E = ± 0.1 TPa σ_{int} = 130 ± 10 GPa at σ_{int} 0.25	AFM	Lee *et al.* 2008
Monolayer Bilayer Trilayer Graphene	E = 1.02 TPa; σ = 130 GPa E = 1.04 TPa; σ = 126 GPa E = 0.98 TPa; σ = 130 GPa	AFM	Li *et al.* 2009

2.3.1 Mechanical Properties

Graphene sheets exhibit high flexibility. They can be stretched like a balloon and even then with stand pressure differences of several atmosphere. They also remains impermeable to even small atoms like helium.

Graphene is a very light material, weighing only about 0.77 mg/m^2. It has tensile high Young Modulus (1.1 TPa), high breaking or fracturing strength (130 GPa) and is the strongest known material (200 times stronger than structural steel). Mechanical properties of the single, bilayer and multiple layers of graphene have been measured by several workers using different methods (Table 2.1).

Park *et al.* (2013) produced a photovoltaic cell using flexible graphene coated with a layer of nanowires. In the future such transparent and flexible solar cells could be deployed on windows and roofs of buildings.

2.3.2 Thermal Properties

Graphene has unmatched thermal conductivity (at 27° C) ≈5000W/ m K^{-1}. Unique thermal properties of graphene are due to strong in-plane carbon bonds; it is an excellent heat conductor. In two-dimensional graphene, there is little or no phonon scattering. In general, the low-energy phonons in the system are involved in heat transfer; hence, it offers higher thermal conductivity.

Graphene exhibits an ambipolar electric field effect such that charge carriers can be tuned continuously between electrons and holes with concentrations as high as 10^{13} cm^{-2} and their mobilities μ in excess of 15,000 cm^2 V^{-1}s^{-1} even under ambient conditions.

The near-room temperature thermal conductivity of graphene is between $(4.84 \pm 0.44) \times 10^3$ to $(5.30 \pm 0.48) \times 10^3$ $W \cdot m^{-1} \cdot K^{-1}$ (Balandin *et al.* 2008). CVD-grown graphene indicates lower values (≈2500W/mK) (Cai *et al.* 2010). It has been suggested that it is the type of structure, i.e., AA or AB type, and number of layers present in the graphene that controls its thermal conductivity as well. Because of its high thermal conductivity, graphene finds its application in electronic circuits as a heat sink. Thermal conductivity values have been obtained by a non-contact optical technique. These values are larger than values obtained with carbon nanotubes or diamond. Thermal conduction is phonon-dominated. The two-dimensional nature of graphene has three acoustic phonon modes. The two in-plane modes have a linear dispersion relation, whereas the out-of-plane mode has a quadratic dispersion relation. Because of this, the T^2 dependent thermal conductivity of the linear mode is dominated at low temperatures by the $T^{1.5}$ contribution of the out-of-plane mode.

Thermal management is one of the key factors in the reliable performance of electronic devices where a considerable amount of heat is generated during operation of the device. Because graphene is being envisioned as a major component in electronic devices because of its high thermal conductivity (up to 5000 W/mK at room temperature, for the single layer of pure defect-free graphene (Ballandin *et al.* 2008), due to strong CAC covalent bonds and phonon scattering. The thermal conductivity of pure single-layer graphene is much higher than the previously reported thermal conductivity of other carbon allotropes at room temperature, e.g., carbon nanotubes (3000 W/mK for MWCNT [Kim *et al.* 2001] and 3500 W/mK for SWCNT [Pop *et al.* 2005]). However, conductivity of supported graphene on various supports varies and needs many studies. The thermal conductivity is affected by factors such as defects, edge scattering (Nika *et al.* 2009) and isotopic doping (Jiang *et al.* 2010). In general, all these factors are detrimental to the conductivity due to phonon scattering at defect and phonons modes localization due to the doping.

2.3.3 Optical Properties

Graphene is known to be transparent (Figure 2.12), which is why it has found applications in many photonic devices that require conducting but transparent thin films. Mono-layer graphene absorbs 2.3% $\pi\alpha \approx 2.3\%$ of white light (97.7 % transmittance), where $\alpha \sim 1/37$ is the fine-structure constant. Stacking order and orientation influence the optical properties of graphene; hence, bilayer graphene shows interesting novel optical properties.

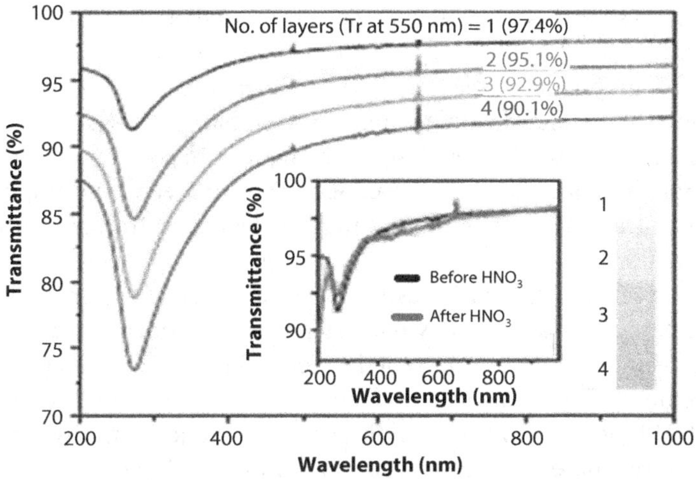

Figure 2.12 %Transmittance by graphene having different layers on quartz substrates. The inset shows the UV spectra of graphene films with and without HNO_3 doping. The right inset shows optical images for the corresponding number of transferred layers (1 X 1 cm^2). Reproduced with permission from Copyright © 2010, Rights Managed by Nature Publishing Group Bae *et al. Nat. Nanotechnol*, 2010, 5, 574–578.

Graphene can be imaged by optical image contrast on Si/SiO_2 substrate due to interference, and the contrast increases with the number of layers. The absorption for monolayer graphene is flat from 300 to 2500 nm, and the peak at ~250 nm in UV region is attributed to the inter band electronic transition from the unoccupied π-states (Figure 2.12).

Along with high transparency, graphene has low resistivity ($10^{-6}\Omega$cm), making it suitable for electrodes in liquid crystal devices.

There is increasing market demand for flexible transparent conductors in touch screens, flexible displays, printable electronics, solid-state lighting, especially organic light emitting diodes, and thin film photovoltaics. Touch screens with transparent windows are seen in smart phones, ATM machines and portable entertainment devices. These demands are being met by indium tin oxide (ITO). ITO film is transparent to the extent of 90% up to a wavelength of 550 nm, with sheet resistances of 10–30 Ω/square.

The problem with using ITO films is that Indium is expensive and scarce. The cost of Indium is about $1000/kg. In addition, this element is being depleted very quickly. Preparative methods, e.g., sputtering, evaporation, pulsed laser deposition and electroplating, are expensive. ITO is a brittle, crystalline material and gets fractured easily. These problems associated with ITO film have made the scientists look for alternatives. In this regard, graphene seems to be the answer to these problems. Graphene is lightweight, robust, flexible, chemically stable and low cost. Graphene sheets have been used for touch screens. In particular, graphene's mechanical strength and flexibility are advantageous compared to indium tin oxide, which is brittle, and graphene films may be deposited from a solution over a large area.

Kalita *et al.* (2010) have measured the transparency of graphene layers and compared its transparency with ITO glass. Graphene layers were synthesized by pyrolyzing camphor of different quantity. Optical and electrical measurement showed a sheet resistance of 860 Ω/sq cm with a transmittance of 91% at 550 nm wavelength of the graphene film. Moreover, more than 80% transparency is observed in the range of 250 nm to 1750 nm; as compared to ITO glass from 250 nm–800 nm. The inset is the photograph of a finger seen through a graphene layer (Figure 2.13).

ITO is used as conductor for the electron/hole in many applications because it is highly transparent (90%) to visible light, has low sheet resistance of 10–30 Ω/square and work function (4.5–5.2 eV). But ITO films have many limitations. This film cannot be used to deposit semiconducting material needing a high temperature, unless ITO film is prepared on

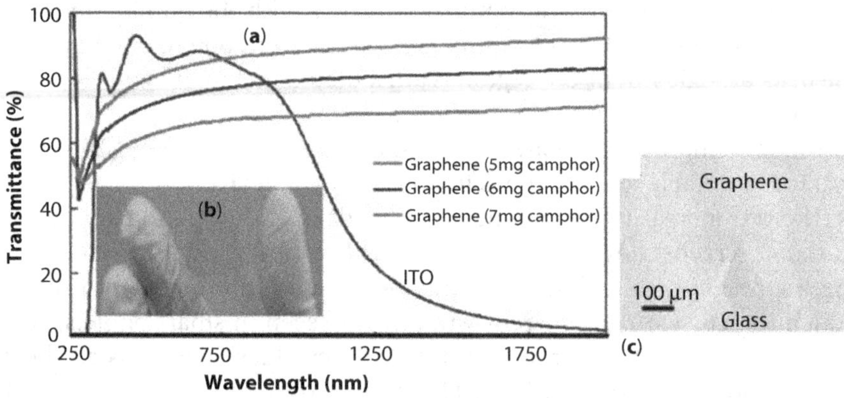

Figure 2.13 (a) %Transmittance by the large area graphene sheets synthesized using different amounts of camphor and transferred on glass. (Copyright © 2010 Elsevier B.V. All rights reserved Courtesy: Kalita *et al.*, *Materials Letters* 64, 2010, 2180–2183) (b) A photograph of monolayer graphene film transferred to PET substrates. (Copyright © 2011 Elsevier B.V. All rights reserved. Courtesy: Kalita *et al.*, *Physica* E 43, 2011, 1490–1493) (c) Large area, few-layer graphene films on arbitrary substrates by chemical vapor deposition. Copyright ©ACS 2009, Courtesy: Reina *et al.*, *Nano Lett.*, Vol. 9, No. 1, 200.

quartz glass, which makes it very expensive. ITO films are unstable in some specific pH levels, which limits its use. Graphene film can be deposited on some metal like copper or nickel, and in situ semiconducting material can be deposited over the graphene layer. Thus it is possible to replace ITO film with graphene film.

2.3.4 Chemical Stability and Reactivity

The high chemical stability of graphene is due to the honeycomb network where strong in-plane sp^2 hybrid bonds exist. The chemically inert nature of graphene has found applications in preventing metal and metal alloys from oxidation. The oxidation resistance of graphene was first shown by Chen *et al.* (2011), when they coated Cu and Cu/Ni with graphene using the CVD technique. Being chemically stable and inert, it is expected to improve the durability of potential optoelectronic devices (Blake *et al.* 2008).

Graphene oxide is soluble in polar and non-polar solvents. Because each graphene atom is present on the surface, these atoms are capable of interacting with any molecule of the target gas or vapor species. Because of this, graphene demonstrates a very special property of any chemical reaction. This property gives the opportunity to adjust the conductivity of graphene by suitably selecting a molecule to adsorb at the surface of graphene. These properties could lead to using graphene for NEMS applications, such as pressure sensors and resonators.

Zhou and Bongiorno (2013), in a recent publication regarding the origin of the chemical and kinetic stability of graphene oxide, have written that:

> At moderate temperatures (# 706° C), thermal reduction of graphene oxide is inefficient and after its synthesis the material enters in a metastable state. Here, first-principles and statistical calculations are used to investigate both the low-temperature processes leading to decomposition of graphene oxide and the role of ageing on the structure and stability of this material. Our study shows that the key factor underlying the stability of graphene oxide is the tendency of the oxygen functionalities to agglomerate and form highly oxidized domains surrounded by areas of pristine graphene. Within the agglomerates of functional groups, the primary decomposition reactions are hindered by both geometrical and energetic factors. The number of reacting sites is reduced by the occurrence of local order in the oxidized domains, and due to the close packing of the oxygen functionalities, the decomposition reactions become—on average—endothermic by more than 0.6 eV.

However, it must be mentioned here that in graphene each atom is exposed for a chemical reaction from two sides due to the two-dimensional structure. Moreover, a carbon atom at the edge of a graphene sheet displays special chemical reactivity, while defects within the sheet also increase the chemical reactivity. The reactivity as well as the electronic properties of graphene depend on the number of layers and varies with the number of layers and the relative position of atoms in adjacent layers (stacking order).

Stacking order in bilayer graphene can be either AA, with each atom on top of another atom; or AB, where a set of atoms in the second layer sits on top of the empty center of a hexagon in the first layer (Figure 2.14D). As the number of layers increases, the stacking order can become more complicated. Carbon atoms of monolayer graphene will be fully satisfied with its valence state, and thus will not create dangling bonds (Figure 2.14A). This will be the case with AA-type structure of monolayer. However, carbon atoms at the surface of the bilayer (of either AA or AB structure) will show some weak dangling bonds due to presence of van der Waal force. This is because one layer below the surface layer is separated by a distance of 0.32 nm and the force responsible to keep the second layer attached with the top surface layer is van der Waal force (Figure 2.14B and D). Because van der Waal force is absent with top carbon atoms, the electrons on the surface of top layers are present in different environments than the carbon of the second layer. As a result, electron interactions with the top layer (Figure 2.14B) would behave differently to what electrons would experience if there was a bilayer. The effect of such behavior is because in

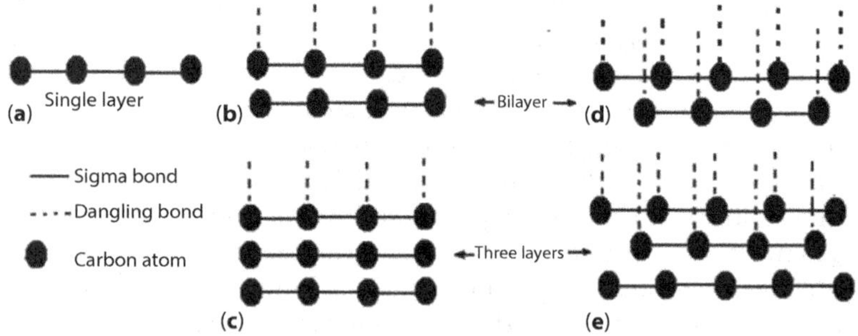

Figure 2.14 (a) Monolayer graphene (b) Bilayer graphene of AA-type structure forming a dangling band with top layer (c) Trilayer graphene of AA-type structure forming a dangling bond effectively with top layer (d) Bilayer graphene of AB-type structure forming a dangling band due to first and second layer carbon, and (e) Trilayer graphene forming a dangling bond due to top layer as well as second layer carbon atom.

AA-type structure, each carbon atom is present just below the top layer carbon atom, whereas with AB-type structure, the dangling bond can be experienced by a second layer. As a result, electron interaction behavior with bilayer graphene of AA-type structure would be different to AB-type structure.

The scenario becomes different when it is trilayer graphene with AA-type structure (Figure 2.14C), though the effect of the dangling bond would behave more or less like bilayer graphene, but the magnitude of electron interaction at the top surface of graphene is different from AA-type bilayer graphene. The situation becomes very complex with three layers of AB-type structure (Figure 2.14E). Dangling bonds are influenced by top layer as well as second layer carbon atoms. The combination of this dangling bond's influence on the electron interaction is different from bilayer graphene. Thus, reactivity of graphene very much depends upon the number of layers and type of structure (AA or AB type.)

The presence of the dangling bond promotes the temporary chemical reaction; in other words it helps in the functionalization of graphene. This is one of the reasons that functionalization can be performed easily with bilayer graphene rather than monolayer graphene.

2.3.5 The Intriguing Electronic Properties (Dirac Point)

A novel electronic property of graphene is that it can sustain huge electrical currents. The π-bonds in graphene contribute to the electron conduction of graphene and provide a weak interaction between graphene layers. The charge carriers in graphene can be described by the Dirac rather than the Schrödinger equation. Due to the two equivalent carbon sub-lattices in its honeycomb lattice, cone-like valance and conduction bands intersect at the Fermi level at the K and K_0 points of the Brillouin zone (Figure 2.15).

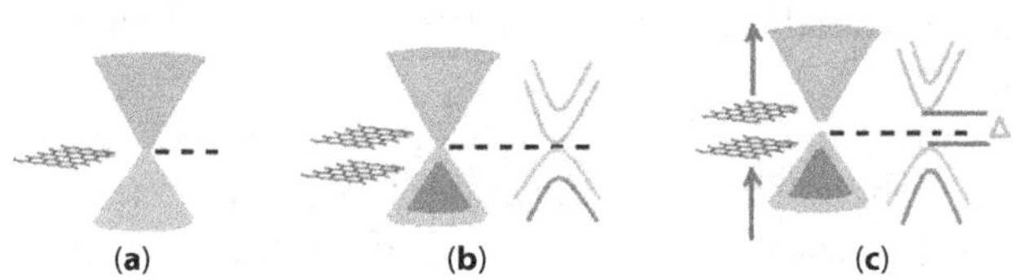

Figure 2.15 (a) Monolayer graphene has a cone-like valance and conduction bands intersect at the Fermi level with no band gap (b) Bilayer graphene with a gapless semiconductor, and (c) Bilayer graphene shows band gap under the electric field when applied in a direction perpendicular to σ-bond.

These massless Dirac fermions reveal many exceptional properties. Graphene being a zero-band gap two-dimensinoal semiconductor, it shows a well-defined ambipolar electric field effect, quasi particles, and large mean-free-path (of the order of micrometers). Furthermore, the Dirac energy dispersion in two dimensions implies that graphene is a gapless semiconductor, whose density of states vanishes linearly when approaching the Fermi energy. Graphene conducts either by electron or hole with concentration as high as 10^{13} cm^{-2}. It shows extraordinary carrier mobility of $\approx$500,000 cm^2 V^{-1} s^{-1}. Such high mobility is because electrons travel through the perfect honeycomb lattice of graphene, resulting in smooth sailing.

Graphene's electronic properties are strongly linked to its thickness. Therefore, it is essential to control the thickness when synthesizing graphene by controlling the growth parameters. Graphene has a high electron (or hole) mobility and also low Johnson noise (electronic noise created due to thermal stir of the charge carriers). This is very useful, especially in making p:n junction so that the dark current is the least.

In 1947, Wallace used a band theory of solids with tight binding approximation to explain many of the physical properties of graphite. According to him, because the spacing of the lattice planes of graphite (which is also true for graphene) is large (3.37Å) compared with the hexagonal spacing in the layer 1.42Å (Figure 2.15), the interactions of electrons between the planes can be neglected and conduction by electrons take place only in layers, i.e., in the π-plane. Later, Novoselov *et al.* (2005) showed that electron transport in graphene can be explained by Dirac's (relativistic) equation. The charge carriers in graphene imitate relativistic particles with zero rest mass with an effective speed of light, $c \approx 10^6$ $cm^{-1}s^{-1}$. As per the characteristic of two-dimensional Dirac fermions, even when the concentration of charge carrier tends to zero, conductivity of graphene does not become zero. The basic difference between the Dirac fermions and Schrödinger equation is that as per the former assumption, electrons propagating through the honeycomb lattice lose their effective mass and results in quasi-particles where as for the latter it is different as shown in Figure 2.15.

There is a drastic variation in the electronic properties of monolayer and bilayer graphene sheet. In the monolayer graphene, the electron is highly mobile and travels with the speed of light. But simply because the graphene monolayer provides highly mobile electron does not make it a very important material unless there is a possibility to control its band gap. Like in semiconductor, one could control and modulate the electron movement to produce desired results. In other words, graphene should not conduct unless energy is supplied to enhance electrons across a gap, i.e., gap between the valence band and the conduction band. Hence if graphene has no band gap, it is like an automobile moving with no control on either speed or turning right or left or stopping due to a traffic light.

Graphene can be made n-type by nitrogen doping using NH_3 or melamine ($C_3N_6H_6$), mixed with PMMA. Nitrogen content is found to be 2–3.5%.

2.3.6 Semiconductor Properties

The interference of the P_z electrons in the honeycomb is responsible for the gapless semiconductor properties, with a point-like Fermi surface. Though graphene is a semi-metal or zero-gap semiconductor, its band gap can be altered by altering the size, shape, thickness and fraction of the sp^2 domains (Kosynkin *et al.* 2009; Li *et al.* 2008).

2.4 Summary

In this chapter, various properties of graphene that make it a different class of material are discussed. The high electrical conductivity, high mobility and high tensile strength are unique properties of graphene, which conducts both electron and holes with almost the same degree of mobility. Moreover, graphene, being almost 90% transparent to visible light makes it more applicable and in that respect it even surpasses the optical properties of ITO glass. Its ability to show chemical reactivity depending upon the number of layers of graphene sheet offers the possibility of several unique applications. Efforts are also made to explain the reasons for possessing such unique properties especially related to the layers of graphene. In spite of all these properties, since the band gap of graphene is either zero that can be increased under electrical fields, this limits its application, especially where band gap plays an important role. However, hydrogenation can increase the band gap (~2.5eV) of graphane and induce magnetic properties in graphone.

3

Nanographene and Carbon Quantum Dots (C-Dots)

The discovery and investigation of the unusual properties of nanoscale materials made of pure carbon were marked by two Nobel prizes. One was awarded for fullerenes, structures of carbon atoms in the form of hollow spheres. Later, fullerenes found their use as scaffolds in designing fluorescence nanocomposites and electron-acceptor quenchers. The other award was for graphene, the two-dimensional sheets of atomic width that have become known as very potent fluorescence quenchers. The emission of their own nanoscale fragments has recently been found and their applications in sensing have been discussed.

The authors

The two forms of graphene that will be discussed in this chapter refer to (i) Nanographene (NG), which also encompasses graphene nanoribbon (GNR) and (ii) Graphene quantum dots or carbon Dots (C-Dots). Material existing in 1–100 nm size is accepted as nanomaterial, whereas quantum dots refer to material having 1–2 nm. Many materials that are up to 5 nm in size and

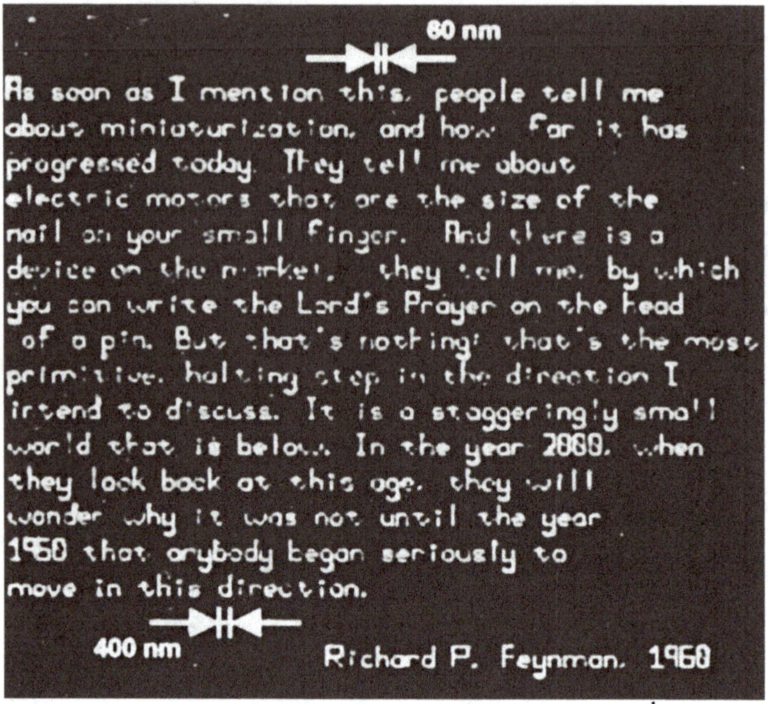

The Founding Speech of Nanotechnology—Written at Nanoscale

exhibit the same properties as quantum dots are also accepted as quantum dots for all the practical and scientific purposes. Moreover, there are many properties that are common or overlapping in nano and quantum dots.

3.1 Nanographene

What is Nano and Nanotechnology? The entry of nanotechnology into modern science took place in the last few decades in a legendary talk entitled "There is Plenty of Room at the Bottom" by Professor Richard Feynman. The brainchild of Feynman was named *nanotechnology* by Taniguchi. Nanotechnology became the subject of the future, the solution of many daunting questions of the century.

Nanomaterials are made from or incorporate particles having a diameter on the order of 1–100 nanometers. At such sizes the material's physical and chemical properties depend sensitively on the particle diameter. It can be defined as:

The design, characterization, production, and application of structures, devices and systems by controlled manipulation of size and shape at the nanometer scale (atomic, molecular, and macromolecular scale) that produces structures, devices, and systems with at least one novel/superior characteristic or property.

The nano-world is not only the world of small dimensions. It is the world in which different materials exhibit new and often unexpected properties. Hence, the race for searching for and synthesizing nano particles of each and every element, polymers and compounds began to be developed, and that led to nanographene.

The inception of nanotechnology is not new! Maharshi Kanad (600 BC), who is considered to be the original proponent of atomic theory, can be considered as the first person to explore the "rooms at the bottom." The concept of nanometals was also instigated by alchemist in the form of Purple of Cassius. But in India, this concept of nanometals is well exploited in ancient *Ayurveda*. Charak Samhita (*A Compendium of Internal Medicine*) and Sushruta Samhita (*A Compendium of Surgery*) are the gifts from Lord Dhanvantari to mankind, and the transmitters of this knowledge were Acharya Charak and Acharya Sushruta, respectively. One of the most important concerns of *Ayurveda* is the noble metal bhasma. It is a matter of adoration that the ancient teachers were not oblivious to nanoparticles in the form of "bhasma."

However, just as Da Vinci's exploration of manned flight has little historical connection to the development of the aviation industry, similarly the nanotechnologists of the medieval era do not really have much to do with the current incarnation of nanotechnology, which can be dated back instead to a talk given by Richard Feynman (1959).

Richard Feynman

Nanographene: Prior to the entry of nanographene (NG) in the nanotechnology and nanoscience arena, many different types of carbon nanomaterials (CNM) were developed, e.g., fullerene, single- and multi-walled carbon nanotubes, carbon nanofibers, carbon nanobeads, carbon nanoonion, carbon nanocones, etc.

Based on their bonding structures, CNM is classified into sp^2-carbon nanomaterials and sp^3-carbon nanomaterials. Typical sp^2-carbon nanomaterials include zero-dimensional (0D) fullerene, one-dimensional (1D) carbon nanotubes (CNTs), and two-dimensional (2D) graphene, all with well-defined structures. The discovery of both fullerene and graphene were awarded the Nobel prize in 1996 (in chemistry to Kroto *et al.*) and 2010 (in physics to Geim and Novoselov), respectively.

Graphene with an ultra-thin nano size (<5 nm) width are called graphene nanoribbons (GNR). Mitsutaka Fujita's group (1996) was the first to introduce a theoretical model of GNR, for examining the edge and nano size effect in graphene (Nakada *et al.* 1996 and Wakabayashi *et al.* 1999).

3.1.1 Structure of Nanographene

Nanographene comprises the similar basic structure as that of graphene as mentioned in Chapter 1, except that nanographene is much smaller (Figure 3.1). However, these smaller fragments of graphene could be spherical, ribbon-like or of asymmetrical shape. With reduced size, graphene displays certain altered and/or enhanced properties.

3.1.2 Properties of Nanographene

Graphene shows interesting inherent optical properties, such as fluorescence. Graphene also has excellent electrical properties, and graphene with all carbon atoms exposed on their surface exhibit ultra-high surface area.

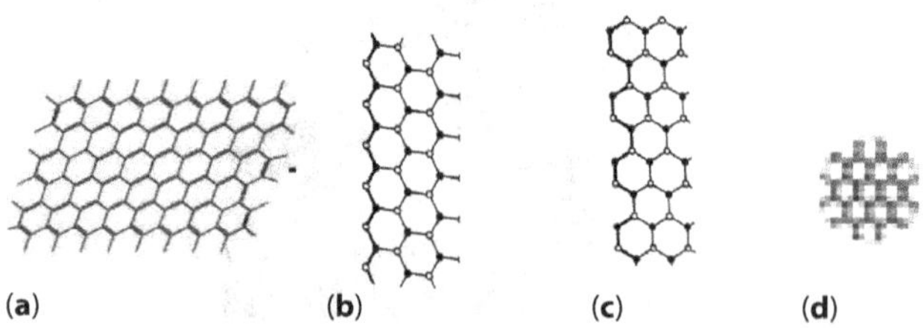

Figure 3.1 (a) Graphene sheet, (b) Graphene nanoribbon with zigzag edge, (c) Graphene nanoribbon with armchair edge and (d) Graphene quantum dots (carbon dots).

With decreasing particle size, the ratio of surface free energy over bulk free energy increases because of the asymmetry in interactions of surface exposed atoms. This means that the energetic, structural and optical properties of nanomaterials become dependent on interactions at their interfaces. In small nanoparticles, most of the atoms are exposed to the surface and participate in interactions with the surroundings. The interactions at interfaces are usually of a different nature from the interactions within the particle core and this may strongly influence the NPs' electronic states and transitions. Excitonic effects in nanographene or C-dots result in strong size-dependent variations in the wavelengths of their emission, so that with increasing size their spectra shift to the red, reaching the near-IR region.

Larger particles, but still smaller than the free path length of an electron, exhibit size-dependent plasmonic effects, while still larger particles exhibit the electronic conductivity of a metal with no emission. The size of nanostructures is usually much smaller than the wavelength of light, a point that is important for understanding a variety of photophysical and photochemical processes occurring on the nanoscale. The electron and proton transfers are short-range phenomena and, if activated in the excited states, they proceed on a scale below 1 nm.

Graphene Nanoribbon (GNR): As mentioned in the previous chapter, the honeycomb structure of graphene shows two orientations; armchair or zigzag at their edges. The electronic properties of GNRs depend on these orientations. GNRs having zigzag edges (Figure 3.1B) are always metallic (as predicted by calculations based on the tight binding, while armchair (Figure 3.1C) can exhibit either metallic or semiconducting behavior, depending on their width.

Density Functional Theory (DFT) calculations show that armchair GNRs are semiconducting with an energy gap scaling with the inverse of the GNR width, i.e., the energy gaps increase with decreasing GNR width. These energy gaps can be opened up to 0.5 eV in a 2.5 nm wide armchair GNR. Zigzag GNRs are also semiconducting and present spin polarized edges. Their gap opens thanks to an unusual anti-ferromagnetic coupling between the magnetic moments at opposite edge carbon atoms. This gap size is inversely proportional to the ribbon width (Son *et al.* 2006 and Jung *et al.* 2009) and its behavior can be traced back to the spatial distribution properties of edge-state wave functions, and the mostly local character of the exchange interaction that originates the spin polarization. Therefore, the quantum confinement, inter-edge super exchange, and intra-edge direct exchange interactions in zigzag GNR are important for its magnetism and band gap. The edge magnetic moment and band gap of zigzag GNR are reversely proportional to the electron/hole concentration, and they also can be memorably controlled by alkaline adatoms (Huang *et al.* 2013).

GNR exhibits high mechanical strength. To improve the mechanical properties of polymeric epoxy nanocomposites, GNRs as well as oxidized GNRs have been used as fillers (Raifee *et al.* 2010). Moreover, graphene displays excellent electron transport properties, as revealed by a number of studies.

3.1.3 Fabrication of Nanographene

Graphene sheet and nanographene fabrication follow almost a similar procedure, but further sequestration of graphene sheet is done by specific method. However, there are reports of directly synthesizing nanographene from precursors, for example, unzipping of CNTs by chemical or plasma etching methods to fabricate GNR. Physical methods have been mostly applied for nanographene synthesis.

3.1.3.1 Physical Methods

1. **Mechanical treatment** of highly oriented graphite is being used to prepare graphene sheets as well as GNR.
2. **Using scanning tunneling microscopic (STM) lithography** is another novel technique to fabricate (Tapasztó *et al.* 2008) GNR with controlled edge orientation. By this method, GNR patterning inside and junctions with nanometer-precision, well-defined widths and predetermined crystallographic orientations and electronic structure can be engineered.
3. **Nanotomy of graphite** is a mixture of physical and chemical method of producing large quantities of GNR. Beery's group has shown that by this method GNR of controlled width can be produced (Mohanty *et al.* 2012). They have used a sharp diamond knife to cut graphite into nano-blocks, which are exfoliated by sulphuric acid to produce GNRs.

   Using the same method, Xiu-Yun (2013) prepared blocks of natural graphite flake. This process involves purification of flakes of graphite by H_2O_2 that was oxidized and intercalated by nitric acid and acetic acid and heating it with microwave. Intercalation in graphene means introducing molecules or heteroatoms in between the original layers of carbons of graphene without the loss of planarity. The graphene layers either accept electrons from or donate electrons to the intercalated species. Due to intercalation, the graphite flakes result into high multiples of expanded graphite, which on ultrasonic dispersion technology along with centrifugal separation produces graphene nanosheets of 1–3 nm or even more than 10 nm.

4. **Plasma Etching:** In this method free non-aggregated MWCNTs (separated by sonication and centrifugation) dispersed in water is used. These MWCNTs are partly embedded in a polymer. Plasma etching of these MWCNTs results in production of GNR having 10–20 nm width and smooth edges (Jiao *et al.* 2009).
5. GNRs can be grown onto silicon carbide (SiC) substrates using ion implantation followed by vacuum or laser annealing (Tongay *et al.* 2012). This technique allows one to write any pattern onto SiC substrates with 5 nm precision.

3.1.3.2 Chemical Methods

Chemical methods mostly involve oxidative modification or conversion of graphene-like structures into GNR.

Chemical unzipping or cutting and opening of multi-walled carbon nanotubes by treating them with potassium permanganate and sulfuric acid (Kosynkin 2009) is another method to fabricate GNR.

3.1.4 Applications of Nanographene

Many biological and electrical applications of graphene utilize nanographene, GNR or carbon dots. The applications are mentioned in brief in here; but it would be better understood after we have discussed the material in chapter 6, detailing the engineering of graphene's properties and discussing the the use of graphene and graphene composites.

1. Because incorporation of GNR in polymeric composite have been found to increase the mechanical strength; recently Lalwani *et al.* (2013) used oxidized-GNR on biodegradable polymeric nanocomposites of poly (propylene fumarate) at low weight % for fabricating bone tissue engineering applications.
2. Because of the unique two-dimensional structure of GNR, its high electrical and thermal conductivity and low noise, it is being considered as a possible alternative to copper for integrated circuit interconnects.

3. The semiconducting properties of GNR have been exploited to be an alternative to silicon semiconductors that may sustain microprocessor clock speeds in the vicinity of 1 THz (Bullis 2008). Using GNR, < 10 nm wide field-effect transistors having I_{on}/I_{off} ratio $>10^6$ at room temperature have been made (Wang 2008a). These field effect transistors can be exploited as channel material for next-generation devices.

3.2 Graphene Quantum Dots or Carbon Dots

What are Quantum Dots (QDs)?—Semiconductor nanocrystals are also known as quantum dots (QDs). Nanoparticles of semiconductors (quantum dots) were theorized in the 1970s and initially created in the early 1980s. If semiconductor particles are made small enough, quantum effects come into play, which limit the energies at which electrons and holes (the absence of an electron) can exist in the particles. As energy is related to wavelength (or color), this means that the optical properties of the particle can be finely tuned depending on its size. Thus, particles can be made to emit or absorb specific wavelengths (colors) of light, merely by controlling their size. Luminescent semiconductor quantum dots (QDs) are nanometer-sized particles with size-dependent optical and electronic properties. QDs' unique and fascinating optical properties are generally not available for individual molecules or bulk semiconductors solids.

Among many properties that exhibit a dependence upon size in QD, two are of particular importance:

i. A blue shift (increase) of band gap energy, when the nanoparticle diameters are below a particular value, which depends on the type of semiconductor. This is called a quantum confinement effect. This effect allows tuning of the energy gap with changes in the QD size. The band gap energy also depends on the composition of the semiconductors as well as the size.
ii. The second significant characteristic is the observation of discrete, well-separated energy states due to the small number of atoms in QDs compared to the bulk. This leads to the electronic states of each energy level exhibiting wave functions that are more atomic-like (Bera *et al.* 2010).

The mechanism of fluorescence of QDs as explained by Joaquin *et al.* (2011) suggests that QDs show distribution of electronic energy states between a discrete molecule and the bulk semiconductor (Figure 3.2a–c). Bulk semiconductors are characterized by band gap energy (Eg), which depends on the composition, and corresponds to the minimum energy required to excite an electron from the ground-state-valence energy band into the vacant conduction energy band. When an electron on the ground-state-valence band absorbs energy more than Eg, then it is excited to the conduction energy band, originating an exciton (negative electron and positive hole pair). The annihilation of the exciton (e.g., by emission of a photon) is known as radiative recombination. The exciton has a finite nm size defined by the Bohr exciton diameter (aB). If the size of the nm crystal is smaller than the size of the exciton, the charge carriers become spatially confined, and that raises their energy. Under this condition, known as the quantum confinement regime, the optical and electronic properties depend on the nanocrystal size. Chemical etching of the nanomaterial causes the formation of corrosion defects, extended imperfections that are functionally bonding to the surface of the NPs. These predominate in relatively small NPs because they may have one-third to one-half of their atoms at the surface. The presence of surface defects alters the luminescence mechanism, as shown in Figure 3.2d. A large number of atoms are located at or near the surface of the NPs, leading to a preponderance of dangling bonds and defects, besides adsorbed impurities. That results in surface states, which can act as traps or recombination sites generating quite complex light-emission profiles. Alternatively, a surface-defect-based luminescence

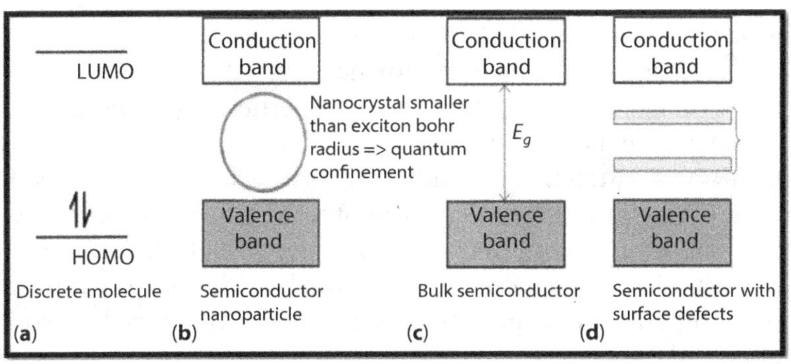

Figure 3.2 Electronic energy states of a semiconductor when it goes from (a) the single molecule to (c) the bulk material, passing through (b) the nanomaterial stage where quantum confinement is observed. (d) The effect of the existence of surface-defect states in the electronic energy state of the bulk material. Adapted from Joaquin C.G., Helena, G. 2011, *Trends in Analytical Chemistry*.

mechanism is suggested for CDs. Supporting this mechanism are results showing that, with different-sized C-dots with similar surface passivation, the smallest were found to be highly luminescent. The presence of surface-defect states near the Fermi level may explain the usually observed dependence of the maximum-emission wavelength and emission-line shape on the excitation wavelength.

3.2.1 Structure of Carbon Dots

Carbon dots (C-dots) are newly discovered nano-engineered materials that represent an important class of carbon nano structures. The size of C-dots is in the range of 5–10 nm (Sun *et al.* 2006). C-dots were accidently discovered in 2004 during the electrophoretic purification of single walled carbon nanotubes (SWCNTs) derived from arc-discharge soot (Xu *et al.* 2004). Since then, much progress has been achieved in the fabrication, properties and applications of C-dots, as recently reviewed by Baker and Baker (2010) and Zhu *et al.* (2009). C-dots possess sp^2/sp^3-disordered carbon or hybridized graphitic core functionalized with polar carboxyl or hydroxyl groups on the surface.

C-dots can exhibit photo-luminescence emission in the near-infrared (NIR) spectral region under NIR light excitation. This property resulted in them being studied in great detail. During the past few years, such tiny C-dots, referred to as fluorescent carbon, have progressed and various methods of synthesis, properties and applications of C-dots, have been achieved and reviewed (Baker and Baker 2010 and Zhu *et al.*2009).

The oxygenous nanoparticles (<10 nm) of C-dots as compared with fluorescent chalcogenide semiconductor nanocrystals (QDs) are promising alternatives due to their excellent optical properties, high chemical stability and low environmental hazard.

The ultrafine dimensions and near spherical geometry (isotropic shape) of C-dots, with their ease of synthesis, has provided a new platform for nano carbon emitters as an alternative to fullerenes or carbon nano tubes. The carbon dots are classified by Salinas-Castillo (2013) in two groups as (i) dispersed or surface functionalized C-dots and (ii) supported C-dots.

However, the mechanism for C-dots is different from the quantum dots of CdSe that has band gap. This means that for CdSe, the semiconducting nature of the material is important, which is not the case with C-dots.

3.2.2 Properties of Carbon Dots

Major unique properties of C-dots are fluorescence, narrow tuneable emission, size-dependent photoluminescence, upconversion photoluminescence, large absorption coefficient and size dependent absorption, broad excitation spectra, multi photon excitation, ease of functionalization, good biocompatibility, good solubility in polar solvents, photostable against photo-bleaching and photo-blinking, and low toxicity. C-dots exhibit nanosecond lifetimes and two-photonic cross-sections similar to those of quantum dots. Technology to form carbon dot shells on different NPs has been suggested (Abhilash, 2010).

3.2.2.1 Optical Properties

a. ***Fluorescence or Photoluminescence***
   When a fluorescent material absorbs light, some electrons move from their ground state into a higher energy excited state. This leaves a vacant spot called a hole where an electron had been. Both the electron and hole can travel, but when they encounter each other again, the electron returns to the ground state, filling the hole and emitting excess energy as light. The color of the light corresponds to the energy difference between the electron's excited state and its ground state; in normal materials, these energy levels are properties of the material's themselves. Therefore, in the normal world, materials that fluoresce red will always glow red and never fluoresce blue. For example, to make a fluorescent lamp that glows a different color, one need to use a different fluorescent material.
   But this is not so at the nanoscale. Quantum dots are so small, typically <10 nm in diameter, that the size of the dot controls the energy levels that electrons can obtain. Therefore, the size and not the material controls what colors are produced.
   Although quantum dots made from cadmium selenide (CdSe) have been extensively researched and are what comes to researchers' minds when quantum dots are mentioned, they may be only one way to obtain quantum fluorescence.
   The discovery of quantum dots made from carbon and previous work on silicon quantum dots are now showing that

there may be more variety in how quantum dots work. The mechanism for carbon quantum dots is different from that of CdSe, which are generally believed to be band gap in origin. This means that for CdSe, the semiconducting nature of the material is important in how light is made, but for carbon, something else is at work. To make carbon quantum dots various methods were used, such as the laser ablation of graphite. These particles formed aggregations and did not fluoresce. The particles, when treated with nitric acid, did not fluorescence. The researchers then coated the particles with simple organic molecules, passivating the surface, which resulted in a bright luminescence (Sheng-Tao Yang 2009).

Luminescence of material exhibits interesting wavelength dependence. Carbon nanodots show a tunable light-emitting range from blue to red under different exciting wavelength irradiations. C-dots show size-dependent photoluminescence and upconversion luminescence properties due to multiphoton process that lead to an anti-Stokes-type emission. The strong and tunable luminescence of carbon materials further enhances their versatile properties; in particular, the quantum effect in carbon is extremely important both fundamentally and technologically (Bourlinos, 2011). The optical behavior may reflect not only effects from particles of different sizes in the sample, but also a distribution of different emissive sites on each C-dot. Mechanistically, it was attributed the PL to the presence of surface energy traps, which become emissive upon surface passivation. It was found that there must be a quantum confinement effect of emissive energy traps on the surface in order for C-dots to exhibit strong PL upon surface passivation (Nagaprasad Puvvada, 2012).

The simple organic molecules, such as diamine-terminated oligomeric poly (ethylene glycol) and poly (propionylethylene-imine-co-ethyleneimine), had no luminescent activity on their own, but when attached to the carbon dots, carbon dots became luminescent. As a result, it was believed that the surface of the carbon quantum dots is source of their luminescent abilities and that the surface passivation is the key to this ability. This may mean that the carbon surface, rather than the entire dot, is the essential part of carbon quantum dots. Surface passivation is somewhat complicated, at least mechanistically, and challenging to pinpoint and understand. Generally speaking, passivation changes the surface, such as the surface terminal atoms, becoming different after attachment of organic species (Bourlinos 2008).

Moreover, the photon emission does not depend on the material quantum dots are made from; it depends upon its size. Generally, the smaller the size of the crystal, the larger the band gap, and the greater the difference between highest valence band and lowest conduction band. Hence, more energy is needed to excite the dots, so more energy released when it returns to ground state. From the above statement, it follows that after excitation, the higher the frequency of light emitted, the smaller the size of the dots will be. The resulting color shifts from red to blue. A most unique property of the Q-dots is quantum confinement, which modifies the DOS near the band edges. Among many properties that exhibit a dependence upon size in Q-dots, two are of particular importance. The first is a blue shift (increase) of band gap energy when the nanoparticle diameters are below a particular value that depends on the type of semiconductor. This is called a quantum confinement effect. This effect allows for tuning of the energy gap with changes in the Q-dot size. The band gap energy also depends on the composition of the semiconductors as well as the size. Second significant characteristic is the observation of discrete, well-separated energy states due to the small number of atoms in Q-dots compared to the bulk. This leads to the electronic states of each energy level exhibiting wave functions that are more atomic-like (Bera *et al.* 2010).

Mechanistically, the fluorescence emissions in carbon dots are attributed to a radiative recombination of the carbon particle surface-trapped electrons and holes, where the large surface (relative to the particle volume) and diverse surface energy trapping sites in the small_ carbon nanoparticles are stabilized by the surface passivation agents. The photoluminescence of C-dots is due to a quantum effect along with different emissive traps on the C-dots' surface.

Photoluminescence is one of the most fascinating features of C-dots. Studies of the optical properties of small-sized C-dots are controversial due to the exact mechanisms of photoluminescence, which remain unsettled and require further clarification. One interesting feature of the photoluminescence of C-dots is the dependence of the emission wavelength and intensity, whether it is due to differently sized nanoparticles (quantum effect) and/or different emissive traps on the C-dot surface, or a mechanism currently unresolved.

The optical behavior need not be due to particles of different sizes in the sample, but may also be due to the distribution of different emissive sites on each C-dot. Li *et al.* (2012) have shown that the C-dots prepared by MAH method exhibited deep UV emission of 4.1 eV (which was the shortest emission wavelength among all the solution-based quantum dots). The mechanisms of C-dot luminescence are not yet clear. However, it has been suggested that it may arise from excitons of carbon, emissive traps quantum-confinement effect, aromatic structures/oxygen-containing groups, free zigzag sites and/or edge defects.

The photoluminescence from C-dots can be quenched efficiently by either electron acceptor or electron donor molecules in a solution, indicating that photoexcited C-dots are excellent electron donors and electron acceptors.

The *size-dependent fluorescence* property of C-dots can be tuned simply by nano-crystal growth, as per desired application.

b. ***Multi-photon excitation***

When multiple low-energy photons arrive "simultaneously" at a fluorophore and, through interaction with it, provoke excitation of an electron that is normally excited by one higher energy photon, multi-photon excitation (MPE) is observed. C-dots show strong fluorescence when excited by UV and near-UV light. Moreover, CDs show luminescence with two-photon excitation in the NIR (800 nm) or long-wavelength light (500–1000 nm).

c. ***Absorbance***
C-dots show apparent optical absorption in the UV region, with a tail extending to the visible range having an absorption band around 220–320 nm (Li *et al.* 2012). However, after surface passivation with different chemical moieties, the absorbance of C-dots was found to increase to longer wavelength. This is due to the presence of surface emissive traps or due to quantum confinement effect. Since C-dots have sp^2/sp^3 hybridized carbon symmetry along with distorted carbon structures suggesting that C-dots are conjugated systems, that might be the reasons for their absorption in the UV-Vis spectral region.

d. ***Upconversion photoluminescence***
Upconversion photoluminescence (UCPL) is due to the electron-accepting and donor property of C-dots, and also due to multi-photon excitation. To explain how UCPL comes from the C-dots, an energy level structural model was established (Li *et al.* 2012), in which the UCPL spectrum could be regarded as an anti-Stokes transition as demonstrated in Figure 3.3d. The energy levels of p and s orbitals are provided by the carbene ground-state multiplicity, which is related to the energy difference (dE) between the p and s orbitals. In this work, the energy between the excitation light and the emission light in the upconversion process was close to 1.1 eV (Hoffmann determined dE which should be below 1.5 eV). When a bunch of low-energy photons excite the electrons in the p orbital, the p electrons will excite to a

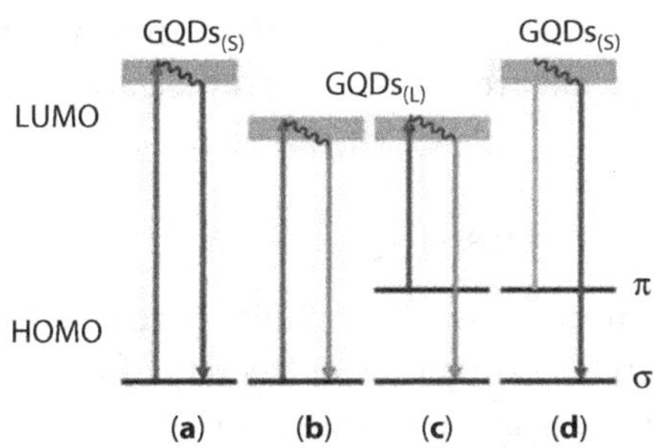

Figure 3.3 Upconversion of photoluminescence of graphene quantum dots (GQD). Copyright © 2012, Royal Society of Chemistry Courtesy Haitao Li *et al. Journal of Material Chemistry* 2012.

high-energy state, e.g., LUMO, and then will go down to a low-energy state. Hence, when the electron transitions back to the s-orbital, an UCPL is emitted. The electrons in the s orbital can also be excited, but they only can emit normal PL (Figure 3.3a). This explains why the difference between upconversion excitation and emission light is a constant energy related to the energy difference of the p and s orbitals.

e. ***Lack of Blinking***

   C-dots do not blink. Blinking is not intrinsically bad, but in some applications, the tendency of some quantum dots to blink on and off is a problem. It was speculated that non-blinking in carbon dots is probably a result of the different luminescence mechanism. This property of C-dots will no doubt lead to more research for applications, since these dots have the potential to be used in medicine where cadmium-based ones cannot (Demchenko, 2013). C-dots' different mechanism of luminescence may already bring about an advantage over traditional types of quantum dots.

3.2.2.2 Photocatalytic Properties

Very pure, high quality, crystalline C-dots having excellent aqueous dispersibility, and remarkable down- and up-converted PL properties, have been demonstrated to possess high photocatalytic activity under visible irradiation, and potential for high efficiency complex catalyst design. Li (2010) demonstrated that their strong and tuneable photoluminescence enables the C-dots to efficiently utilize the full spectrum of sunlight. Interestingly, the PL from C-dots can be quenched efficiently by either electron acceptor or electron donor molecules in solution, indicating that photoexcited C-dots are excellent electron donors and electron acceptors (Salinas-Castillo, 2013).

3.2.2.3 Chemical Inertness

Chemical inertness means no or little ability to react with other elements or compounds; it depends mainly on the valency of elements. Elements usually react with other substances because they have to satisfy their valency by filling their octets, but those elements which already have a completely filled octet shows no or little reactivity towards other elements and rarely react with other elements. This is known as chemical inertness. C-dots are chemically inert, and hence are stable, due to which they can be used in various applications.

3.2.2.4 *Water Solubility*

C-dots exhibit inherently low and amphiphilic characteristics depending on the surface-capping materials (Bourlinos 2008). In addition, typical C-dots contain a large amount of carboxylic and hydroxyl groups at their surfaces, thus endowing them with excellent water solubility and the suitability for subsequent functionalization with various organic, polymeric, inorganic or biological species.

3.2.3 Fabrication of Carbon Dots

The synthesis methods for C-dots can be divided into two categories: top-down and bottom-up routes (Hsu and Chang, 2012).

The *top-down* routes consist of delaminating bulk graphite or carbon materials into nanosize graphite particles via physical or chemical methods. These include arc discharge, laser ablation, electrochemical oxidation and chemical oxidation, and so on. However, there is no fluorescence from these nanosized graphite particles. Another passivation step is necessary to produce light emission, which is associated with defects in the graphene structures (Haitao Li *et al.* 2012).

Bottom-up methods are approaches that convert non-graphite carbon sources into nanosize graphite particles via chemical reactions, for example, combustion, pyrolysis, microwave, ultrasonic, hydrothermal methods (Qinlong Wang 2011).

A variety of carbon precursors, such as graphene sheets, sucrose, amino acids, chitosan, and different surface passivation agents such glutathione, polymers like PLGA, and PEG have been used for the synthesis of carbon dots or structural or surface functionalization. It should be noted that NIR PL emission of C-dots excited by NIR excitation is particularly significant and useful for in vivo bionanotechnology because of the transparency of body tissues in the NIR "water window." An important finding of carbon dots is that, unlike the semiconductor QDs containing heavy metals such as cadmium, carbon is generally not considered to be a toxic element. An additional recent finding reveals that C-dots can exhibit PL emission in the near-infrared (NIR) spectral region under NIR light excitation. This property of NIR PL emission of C-dots excited by NIR excitation is particularly significant and useful for in vivo bionanotechnology because of the transparency of body tissues in the NIR "water window."

A number of methodologies to produce C-dots have been reported. These metholodogies include chemical and physical methods. Examples are listed below.

3.2.3.1 *Electrochemical Methods*

Various electrochemical methods of synthesis of C-dots have been reported. Some of them are mentioned below:

1. Oxidizing a graphitic column electrode against a saturated calomel electrode with a Pt wire counter electrode in NaH_2PO_4 aqueous solution (Zhou J. *et al.*, 2007).
2. Using an assembly of a graphite rod working electrode, a Pt mesh counter electrode, and an Ag/AgCl reference electrode immersed in pH 7 phosphate buffer solutions (Zheng L. *et al.*, 2009)
3. Ionic liquid (IL)-assisted electro-oxidation of graphite using the water soluble IL 1-butyl-3-methylimi-dazolium tetrafluoroborate [BF_4] that has 90 wt% water as electrolytes. (Xu *et al.*, 2013).
4. Alkali-assisted electrochemical method is also used for generating 1–4 nm C-dots (Li *et al.* 2012). In this method, a graphite rod is used as both anode and cathode, and NaOH/EtOH as electrolyte with a current intensity of 10–200 mA cm^{-2} (Figure 3.4). Graphite honeycomb layer is cut into ultra small particles that lead to fragmentation of graphite into tiny, high quality C-dots. It must be mentioned here that when sulphuric acid (H_2SO_4/EtOH) was used as an

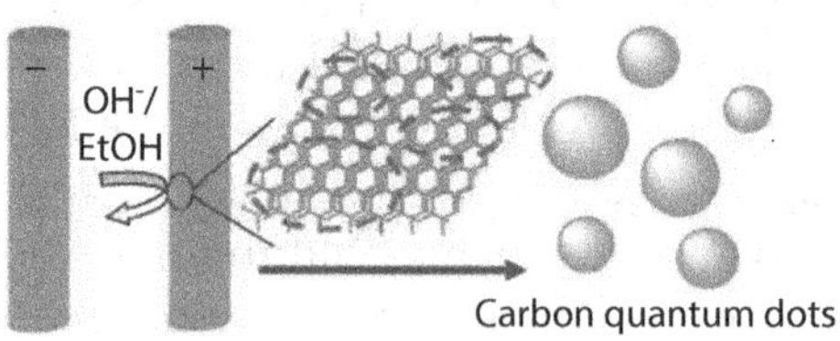

Figure 3.4 The schematic diagram of electrochemical fabrication of C-dots. Courtesy Copyright © 2012, RSC publishing,: Hiatao Li *et al.*, *J.Material Chemistry*, 2012,22, 24230-24253 DOI: 10.1039/c2jm34690.

electrolyte, C-dots did not form, thus confirming that an alkaline environment and OH groups are essential for the formation of C-dots by electrochemical oxidation process. For large-scale synthesis of high-quality crystalline C-dots with high purity, a method is proposed by Ming *et al.* (2012), where pure water is used as an electrolyte without any other chemical additives (Figure 3.5). C-dots thus obtained shows excellent aqueous dispersibility, and remarkable down- and up-converted PL properties. Above all, this method requires no further purification.

5. The electrochemical oxidation of graphene electrode in phosphate buffer solution can also yield 3–5 nm uniform size C-dots of 1–3 graphene layers due to strong inter-graphene attraction (Que *et al.* 2011). The oxygen-containing groups on the surface of C-dots enables aqueous solubility and facilitates surface functionalization.

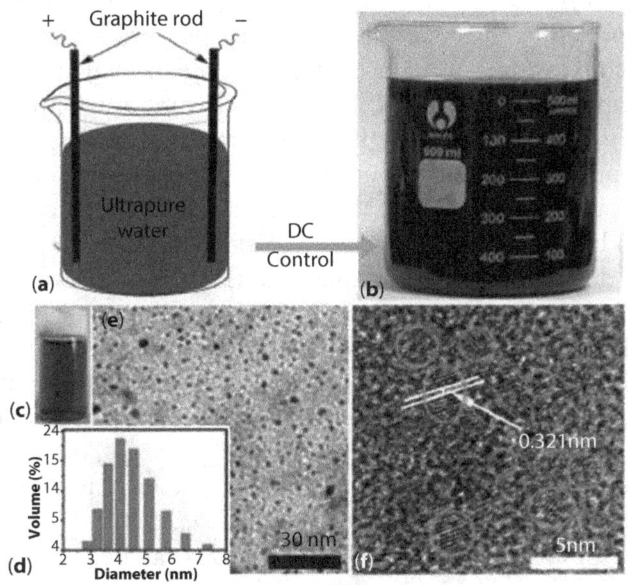

Figure 3.5 (a) Reaction equipment for the preparation of C-dots; digital image of C-dots solution (b) before treatment, (c) after treatment; (d) digital DLS histogram of C-dots; (e) TEM (f) HRTEM image of C-dots. Courtesy Copyright © 2012 RSC Publishing: Zheng Ma *et al.*, *New J. Chem.*, 2012,36, 861–864.

3.2.3.2 Combustion, Thermal, Hydrothermal and Acidic Oxidation of Carbon Precursors

Thermal decomposition methods can directly produce functionalized C-dots, but require a suitable carbon source.

In the combustion oxidation method, (Liu *et al.* in 2007) soot from a burning candle is collected on either aluminium foil or a glass plate and then mixed with oxidant (e.g., nitric acid) and refluxed to oxidize the soot particles present on the surfaces. C-dots thus formed are separated by centrifugation (in a solvent mixture of water-ethanol-chloroform) or dialysis and then subjected to polyacrylamide gel electrophoresis (PAGE) fractionation.

Soot of combusting natural gas and paraffin oil is also a precursor for the fabrication of C-dots. Refluxing the combustion soot is done using nitric acid (Tian *et al.* 2009) to synthesize C-dots. By this method 4.8 ± 0.6 nm C-dots are obtained.

Another thermal oxidation process that has successfully synthesized C-dots is by pyrolysis of ehtylenediamine tetraacetic acid salts, by controlled carbonization of sucrose (Wu *et al.* 2010). The preparation of highly photoluminescent C-dots in MAPO-44, an Mg substituted microporous aluminophosphate molecular sieve with a chabazite structure, through thermal decomposition of the occluded template or loaded organic molecules, was reported by Chen *et al.* The resulting composite phosphors could be excited by a broad range of light in the ultraviolet region, and the emission wavelength is tuneable by varying the thermal treatment condition.

In the solvothermal method, a precursor (Glucose in water-glycol medium) is heated at 180°C in presence of nitric acid to achieve C-dots that are modified with polyethyleneimine (PEI) to improve fluorescence quality.

Hydrothermal oxidation of L ascorbic acid or hydrothermal cutting of graphene sheets (Pan *et al.*) yields 2–3 nm C-dots. Hydrothermal oxidation of nano-diamond has also been reported as a method to fabricate tuneable C-dots (Zhang *et al.* 2013).

A facile one-step wet chemical synthesis of C-dots by acidic oxidation of carbon fibers (CF) has been achieved (Peng *et al.* 2012). Interestingly, the PL from the CF-derived C-dots of different sizes can be tailored by simply choosing different reaction temperatures.

A mixed acid (sulphuric acid: nitric acid 3:1) treatment (refluxing) for oxidation of graphite/CNT also produces nearly identical C-dots of 3–4 nm. For separation of C-dots the standard method of dispersal of mixture in water and then filtration and dialysis is done to remove CNT (Lee *et al.* 2012).

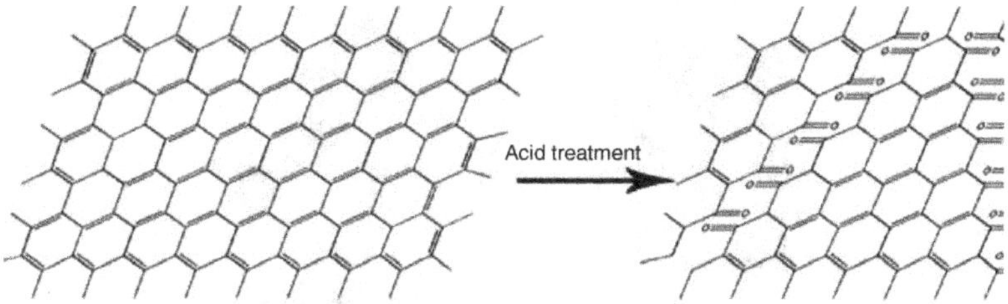

Figure 3.6 Chemical oxidation of carbon fiber into C-dots. Based on results obtained by Peng *et al.* 2012.

SWCNT is also used as precursor. It is first etched by refluxing with nitric acid, which yields C-dots with some disorders, then it is hydrothermally oxidized by suspending it in water and heating it at high temperature (200°C) (Dong *et al.* 2013).

Another more simplified procedure of hydrothermal oxidation of carbon compound to C-dot is by directly heating ascorbic acid aqueous solution at 90°C. The resulting carbon dots were found to possess excitation, pH and polarity-dependent luminescence as well as upconversion fluorescence properties.

Acidic oxidation of micro-sized pitch based carbon fibers that has stacked graphitic zones also produces C-dots. Peng *et al.* (2012) have shown that these C-dots are 1–4 nm in size, 1–2 atomic layers thick, and mostly have zigzag edges (Figure 3.6).

3.2.3.3 Pulsed Laser Irradiation of Carbon Source

The basic difference between graphene quantum dots and carbon dots is that spherical carbon dots are single layer structures, whereas graphene quantum dots (GQDs) are planar nanoflakes a few atomic layers, thick thus having higher surface-to-volume ratio than spherical carbon dots. Pulsed laser irradiation is a good method for synthesizing GQD. Benzene has been successfully used in this method using nickel oxide as catalyst. Both, when irradiated with 1064 nm pulsed Nd:YAG laser, resulted in the formation of GQD. Nickel was separated by centrifugation and the remaining benzene is separated by rotor evaporation. This process gives a good yield of water-soluble 2.6 nm GQD across three layers (Habiba *et al.* 2013).

Laser ablation (Nd:YAG pulsed with a second harmonic wavelength of 532 nm) of nanocarbon dispersed and sonicated in liquid medium (ethanol, acetone or water) also yields good quality C-dots (Li *et al.* 2012). In this process, the suspension of nano-carbon has to be taken into a glass cell for laser irradiation. Figure 3.7 describes the experimental setup.

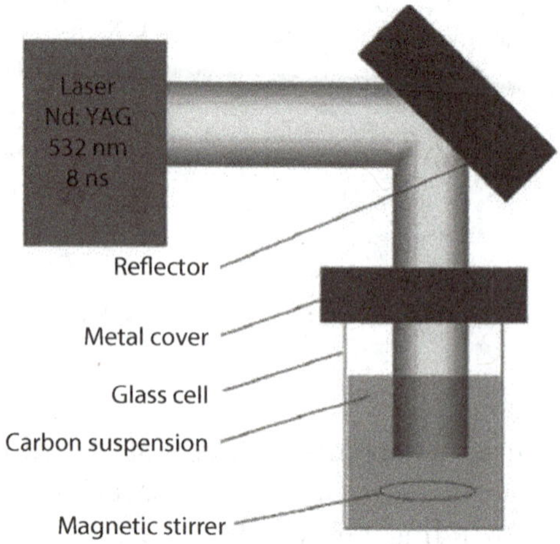

Figure 3.7 Representation of experimental setup of laser ablation. Courtesy Copyright © 2011, RSC publishing: Li X. *et al. Chem. Commun.*, 2011, 47, 932–934.

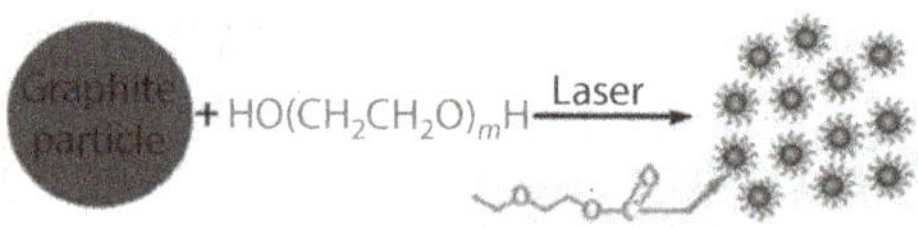

Figure 3.8 One-step synthesis of C-dots in PEG200N solvent. Courtesy Copyright © 2008, Royal Society of Chemistry: Hu S. *et al., J. Mater. Chem.*, 2009,19, 484–488.

3.2.3.4 *Laser Ablation of Graphite*

C-dots are produced by laser ablation of a carbon target in the presence of water vapor with argon as a carrier gas at 900°C and 75 kPa. These C-dots need further refluxing in HNO_3 for up to 12 hr and also require surface passivation by attaching simple organic species. The acid-treated C-dots give bright luminescence emission (Sun *et al.* 2006).

Synthesis of fluorescent C-dots by laser irradiation of a suspension of carbon materials in organic solvent has also been reported (Hu *et al.* 2009) (Figure 3.8).

3.2.3.5 *Arc Discharge Method*

Synthesis of SWCNT by arc discharge method is a well-known process. However, during this process, some fragments of carbon material (soot) are also formed. To separate and purify the C-dots, they are oxidized with nitric acid, and finally the segment is extracted with a NaOH solution (pH 8.4) as a stable black suspension (Xu *et al.* 2004) containing SWCNT and C-dots. C-dots and SWCNT are further separated by gel electrophoresis (Li *et al.*, 2012).

3.2.3.6 Plasma Treatment Method

A one-step combined synthesis and functionalization of C-dots using an all-in-one small submerged arc plasma reactor can be done by taking advantage of the long-lived free radicals generated by submerged-arc helium atmosphere plasma generated on the nanoparticle surfaces to supply ethylnediamine to functionalize the C-dots (Jiang *et al.* 2009).

3.2.3.7 Opening of Fullerene Cage

This is a mechanistic approach for the synthesis of C-dots developed by Loh *et al.* (2010) by metal-catalyzed cage opening of C_{60} fullerene. The fragmentation of the embedded molecules at elevated temperatures produces carbon clusters that undergo diffusion and aggregation to form C-dots.

3.2.3.8 Ultrasonic-/Microwave-Assisted Synthesis

Microwave-assisted techniques have shown significant progress in their application to materials synthesis because of rapid heating and the resulting dramatic increase in reaction rate. C-dots produced by a microwave method showed bright luminescence when their surfaces were passivated with diamine-terminated oligomeric poly(ethylene glycol) (Zhu *et al.* 2009). An additional advantage is that C-dots are produced in a significantly shorter time. The excellent heating capacity of microwave irradiation lies in the fact that it can heat substances uniformly. It is believed that microwave irradiation produces a force on charged particles and causes them to move or become further polarized. This leads to friction and collision of molecules, which subsequently generates heat. Rapid heating by microwave irradiation eliminates the effect of a temperature gradient, contributing to the mono-dispersion, small size and high purity of the resulting C-dots. The microwave-assisted pyrolysis approach to synthesize fluorescent C-dots with electro-chemi-luminescence (ECL) properties is a rather cheap process. In this process, solution containing a carbon source (polysaccharides, glycerol, glycol, glucose, sucrose, etc.) is exposed to microwaves (Wang *et al.* 2007). The reaction time is only few minutes, and C-dots can be produced without any passivating agent. With increasing reaction time, the solution changed from colorless to yellow, and finally to dark brown, which implied the formation of C-dots.

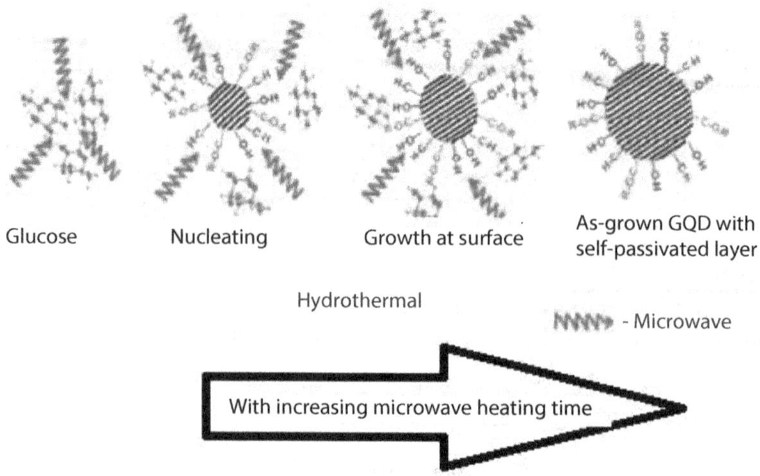

Figure 3.9 Preparation of GQDs by microwave-assisted hydrothermal method. Courtesy Copyright ©2012, American Chemical Society: Tang L. *et al.*, *ACS Nano*, 6, 5102, 2012.

A microwave-assisted method in combination with hydrothermal method using glucose can also produce water-soluble crystalline C-dots with an average diameter as small as 1.65 nm. This method does not require any surface passivation agents or inorganic additives. C-dots can be produced just by pyrolyzing glucose, sucrose or fructose (Figure 3.9) (Lee *et al.*, 2012).

C-dots can also be fabricated by the ultra-sonication of glucose or active carbon. Such monodisperse water-soluble fluorescent C-dots were found to emit bright and colorful PL, covering the entire visible-to-NIR spectral range.

Some research is also being done to create quantum dots by changing the width of GNRs at select points along the ribbon, creating quantum confinement (Wang *et al.* 2007).

3.2.3.9 *Chemical Methods*

For chemical approaches, solution phase chemical methods by oxidative condensation of aryl groups have been successfully applied to produce C-dots. Versatile synthesis of large, stable colloidal C-dots with desired sizes and structures via a new solubilization strategy was demonstrated by Li *et al.* (2012). Recently, unsubstituted hexa-perihexabenzocoronene (HBC) as a carbon source was used to synthesize multicolor C-dots with a uniform size of 60 nm diameter and 2–3 nm thickness. This method has

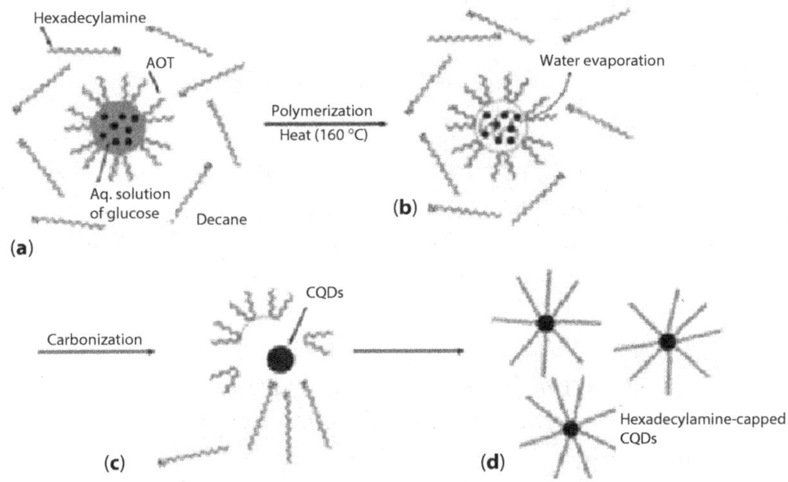

Figure 3.10 Schematic representation of synthesis of C-dots in reverse micelles: (a) formation of water-in-oil reverse micelles, (b) hydrolytic polymerization, (c) carbonization and (d) in situ passivation by hex-adecylamine. Courtesy Copyright © 2012, American Chemical Society,:Kwon *et al. Chemical Commun.* 2012, 48, 5256–5258.

produced the largest C-dots reported so far. A method employing reverse micelles as nanoreactors is also used to produce highly luminescent C-dots via glucose carbonization in reverse micelles followed by in situ surface passivation, as shown in Figure 3.10. This solution-phase chemical method offers size tenability and narrows size distribution without any impractical size separation process (Lee S. *et al.*, 2012).

3.2.3.10 *Supported Synthetic Procedure*

Supported synthetic procedure has been widely adopted for the synthesis of monodisperse nanomaterials, involving molecular sieve, porous carbon and so on. This method has been used to produce C-dots as well. For C-dot synthesis, various supported synthesis routes have been employed as discussed below.

i. Li *et al.* used surfactant-modified silica spheres as supports to localize the growth of C-dots by blocking nanoparticles agglomeration during high-temperature treatment.
ii. Thermal oxidation of an appropriately ion-exchanged NaY zeolite is also used for the synthesis of supported C-dots (4–6 nm).

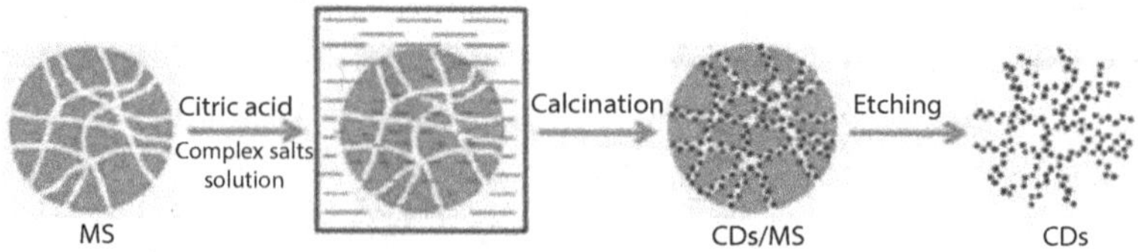

Figure 3.11 Processing diagram for the synthesis of photoluminescent carbogenic dots. Courtesy Copyright © 2010, Royal Society of Chemistry: Zong J *et al.*, *Chem. Commun.*, 2011, 47, 764.

iii. A facile approach for preparing hydrophilic C-dots is by using mesoporous silica (MS) spheres as nano-reactors in which C-dots are impregnated. The overall synthetic procedure is illustrated in Figure 3.11. First, MS spheres are prepared with N-hexadecylamine as the surfactant and tetraethoxysilane (TEOS) as the precursor using ammonia as a catalyst. Subsequently, MS spheres are impregnated with a mixed solution of complex salts and citric acid. Subsequently, calcinations and removal of MS supports generated the nanosized hydrophilic C-dots. The employment of MS spheres as supports is the key feature, which not only confines the C-dots to a narrow size distribution in the pores of MS spheres, but also prevents the aggregation of the nanosized C-dots.

3.2.3.11 Biogenic Synthesis

Biogenic synthesis of C-dots involves use of natural plant or plant-derived materials as precursors or sources of carbon, e.g., orange juice, jaggery, bread, sugar, sugarcane juice, *Trapa bispinosa* peel extract, etc., to synthesize highly fluorescent C-dots. Sharon's group (Pandey *et al.*, 2013) has successfully synthesized and achieved size-dependent separation by sucrose density gradient centrifugation, luminescent water-soluble C-dots using water plant *Trapa bispinosa* peel extract without adding any external oxidizing agent at 90°C. These C-dots showed prominent green fluorescence under UV light (λmax). UV-Vis spectra of these C-dots were recorded at different time intervals (30–120 nm). The data obtained displayed signature absorption of C-dots between 400 and 600 nm. Fluorescence spectra of the dispersion were recorded after 120 min of synthesis. These spectra exhibited characteristic emission peaks of C-dots when excited at 350, 400, 450 and 500 nm. X-ray diffraction (XRD), Raman spectroscopy and thermogravimetric analysis (TGA) was used to further analyze C-dots. XRD revealed that the structure of the C-dots was turbostatic. C-dots produced by this method were found to be exceptionally biocompatible with MDCK cells.

Hydrophobic C-dots can be synthesized from gum arabic by microwave-assisted methods in the presence of ethanol and NaOH in a domestic microwave oven for 4 min. The C-dots were extracted using chloroform. They were highly dispersed in chloroform as well and the corresponding PL spectrum at different excitation wavelengths is observed with a red shifted pattern (Qinlong Wang, 2011). The excitation wavelength dependent emission wavelength and intensity were attributed to quantum effects and/or emissive traps present on their surface owing to the presence of functional groups. At certain excitation wavelengths, some emissive sites would be excited and fluoresce, giving rise to excitation-dependent emission spectra. Partial self-passivation led to variation in sizes; however, a distinct spherical morphology was maintained throughout. The carbon dots were comprised of –OH groups, C–H groups, carbonyl, C–O and C–O– groups. The carbon plane had poor crystallinity or was nearly amorphous in nature. Although the synthesized CDs appeared as a brownish-black precipitate in water, they were immediately extracted into organic solvents and therefore the hydrophobicity of such CDs was illustrated simply with the ease of dispersion in different organic solvents. However, no significant green coloration was observed from the aqueous layers. The mechanism of formation of hydrophobic C-dots was probably by fragmentation of gum arabic forming the carbon core, followed by mild partial passivation via condensation with hydroxyl groups of fragmented or unfragmented gum under microwave heating conditions forming C–O–C bonds, which is known to occur with active hydroxyl end groups. The remaining marginal hydroxyl and carbonyl groups on the surface of those hydrophobic C-dots resulted in the formation of defect sites and emissive traps, which contributed to its PL properties. Radiative recombination of excitons trapped within the defects produced the most intense PL band and consequently its strong PL. On the other hand, reduced surface hydroxyl groups along with hydrophobic long chains of fragmented or unfragmented gum resulted in its hydrophobic character. This speculated mechanism is significantly similar to the other organophilic carbon dots, where long chain amines contributed to passivation after amide bond formation on the carbon dot. However, our microwave-assisted approach was undoubtedly convenient to produce hydrophobic dots and unique as well in comparison to the methods reported so far. Carbon dots were highly dispersed in a wide variety of organic solvents including acetone, ethanol, chloroform, toluene, THF, NMP, hexane, cyclohexane, DMF and acetonitrile (Bourlinos 2008).

In short, by carbonizing the rich carbon sources and in some cases by passivating the surfaces, photoluminescent and chemically inert carbon dots are synthesized.

3.2.4 Applications of Carbon Dots

Photoluminescent properties, in combination with their broad absorption spectra, large absorption coefficients and good solubility in polar solvents, make the carbon dots suitable as alternatives to semiconductor quantum dots and expensive heavy metals commonly used for photovoltaic and photocatalytic application. Most C-dots show excitation in the UV region (330–420 nm). They also yield photoluminescence in the range of 400–699 nm depending on their size. However, C-dots are found to have emission wavelength that are dependent on their size, and they alone cannot provide high specificity towards a target (Lee S. *et al.*, 2007).

The superiority of C-dots in terms of high aqueous solubility, robust chemical inertness, easy functionalization, high resistance to photobleaching, low toxicity and good biocompatibility enable them to be used in diverse array of biological applications as well.

However, with these magnificent properties, C-dots still need to be studied in great detail to exploit their properties in nanotechnology and nanomedicine. Within a short period, C-dots have demonstrated their latent potential in a variety of applications, such as sensing, biological labeling, photo-catalysis, and optical-electronic devices.

3.2.4.1 Sensor Designing

C-dots have strong intrinsic peroxidase-like activity. Based on this intrinsic peroxidase activity, a simple, cheap, and highly selective and sensitive colorimetric and quantitative assay can be developed for the detection of biomolecule. Based on their excellent properties, C-dots can be designed for sensors, such as biosensors for DNA nitrite sensing, sensors for phosphate, glucose, a-fetoprotein or metal ions. Research studies have reported the use of unmodified C-dots as novel and environmentally friendly fluorescence probes for the sensing of Hg^{2+} and biothiols with high sensitivity and selectivity. C-dot probes are used for fluorescent Hg^{2+} detection application, which is based on Hg^{2+}- induced fluorescence quenching of CPs. This sensing system exhibits excellent sensitivity and selectivity toward Hg^{2+}, and a detection limit as low as 0.23 nM is achieved. The practical use of this system for Hg^{2+} determination in lake water samples has also been demonstrated successfully (Zhen Lin, 2011).

3.2.4.2 Bioimaging

Bioimaging sensors for in vivo bioanalytical diagnostics must be non-toxic and biocompatible. In contrast with quantum dots, which usually contain cadmium, C-dots comprise non-toxic elements, which makes them a particularly useful, promising bioanalytical tool. The first preliminary assessment of the bioimaging potential of C-dots was done using labeled *E. coli* ATCC 25922 cells with the C-dots. The cellular uptake of C-dots by Caco-2 cells was recorded by confocal microscopy.

When C-dots were injected into mice, they remained strongly fluorescent *in vivo,* which, coupled with their biocompatibility and nontoxic characteristics, might offer great potential for optical imaging and related biomedical applications.

Live-cell-imaging was done under confocal microscope with laser excitation of 405 and 488 nm, and fluorescence was collected in blue and green regions, with the help of C-dots on human osteosarcoma cell line MG63.

For subcutaneous injection, female DBA/1 mice, upon injection of a C-dot solution or a CZnS-dot solution, were imaged in a Lumazone FA in an *in vivo* imaging system. The fluorescence images of subcutaneously injected mice exhibited bright emission for both of np's, and the fluorescence faded in around 24 hours (Nagaprasad Puvvada, 2012).

There have been efforts to enhance the bioimaging potential of C-dots using the following methods:

a. Doping it with N_2 or S (Dong *et al.* 2013). However, even without functionalization, C-dots showed their ability to enter cells and Ray *et al.* have used such C-dots in cell imaging (Ray *et al.*, 2009).
b. Surface passivation with branched polymer amplified the radiant efficiency of C-dot, which was used for imaging lymph nodes in mouse models (Wu *et al.* 2013).
c. Huang *et al.* (2012) conjugated C-dots with fluorescent dye, which absorbs in NIR to study its in vivo bio-distribution; by conjugating with dye Ce6, they could suppress tumor growth using photodynamic therapy.
d. Conjugating C-dots with iron oxide, when it was injected in rats, was found in spleen samples when screened in near IR (Srivastava *et al.*, *Small* 2012).

e. Making a C-dot and mesoporous silica complex, Sharon's group used rice husk, which produced highly green fluorescent (under UV-light) C-dots embedded in mesoSiO_2. This complex was used for labeling yeast cells, becuase fungi is a close representative of eukaryotic organisms. Furthermore, a mesoSiO_2/C-dot complex was found to be highly biocompatible for Vero cells (Pandey *et al.* 2013).

There have been different angles through which bioimaging of various cells using C-dots have been studied. Zhang *et al.* (2010) demonstrated polydopamine and cellulose derived C-dots for cellular imaging. Similarly, Wei *et al.* (2010) used paper ash C-dots synthesis and cellular localization of L02 hepatic cells. Chen *et al.* modified sucrose-derived C-dots with oil for imaging 16HBE cells, whereas Zhang *et al.* (2010) used BSA and TTDDA for viewing colorectal cancer cells.

3.2.4.3 Drug Delivery

Semiconductor quantum dots of CdTe, CdS, CdSe, etc., have been vastly used in bio-imaging and as biological sensors due to their distinct optical properties and small size. However, being heavy metals, they have been cytotoxic; hence the use of QDs of heavy metals for practical applications in the biological cells has been restricted. With the advent of C-dots having many cyto favorable properties, scientists are exploiting C-dots for drug and nutrient delivery. C-dots' properties that deem them to be a suitable drug delivery vehicle are:

a. ***Biocompatibility:*** Carbon, being an inert material, does not cause any toxic effects on cells and can be used for in vivo biological analysis. So far, there are no reports for toxicity of C-dots on cells or mice.
b. ***Tuneable optical properties:*** C-dots have drawn great attention in the field of research due to their excellent tuneable optical properties. It is a characteristic of QDs and nanoparticles in contrast to bulk materials that allows for designing size-dependent photo-physical properties as per the desirable application.
c. ***Solubility in water:*** For applications like drug delivery and bioimaging, water-soluble C-dots hold an added advantage over conventional quantum dots. They allow easy harvesting and functionalization of C-dots with drugs and other vital

molecules, and also reduce toxicity caused by other inorganic solvents. Moreover, synthesis, purification and storage also become simple.

d. ***Resistance to photobleaching:*** There have been many reports of the fluorescence emission of C-dots, stating that they are comparatively resistant to photobleaching. This is also a generally observed trait of quantum dots, which consequently leads to their blinking nature and hinders accurate detection for sensor purposes due to weak signals. On the other hand, C-dots are stable in fluorescence emission for long hours without alteration in their PL spectra.

C-dots have been found to be localized not only in the cytoplasm (Mewada *et al.* 2013), but also in the cell membrane (Zhang *et al.*, 2012).

Sharon's group showed specific targeting potential of C-dots using antibodies and/or targeting ligands in cancer, for neurodegenerative disease, to deliver antibiotics and C-dots functionalized gold nanorod as tri-functional nano-worms i.e. for cancer drug delivery, photothermal therapy and bioimaging (Mewada *et al.* 2014, Pandey *et al.* 2013a, Thakur *et al.* 2014 and Pandey *et al.* 2013b).

C-dots derived from Sorbitol can be used for the targeted delivery of Doxorubicin, an anti-cancer drug using folic acid (FA) as a navigational molecule. Folic acid is highly expressed in most of the cancer cells; hence, it is a suitable molecule for targeted delivery of anti-cancer drugs. However, prior to the attachment of Doxorubicin, the surfaces of C-dots need to be protected with bovine serum albumin to make them more biocompatible and to attach and hold high amounts of drugs. Mewada *et al.* (2013) have shown the internalization of C-dots specifically targeted with FA in HeLa cells.

Recently, a unique blend of C-dots and gold nanorods was prepared. This complex was used for anchoring Doxorubicin via covalent and non-covalent pH-sensitive chemical bonds and tried for photothermal therapy as well as the delivery of therapeutic payloads because of their typical size and shape-related features. These C-dots and gold nanorods have been found to be suitable for the controlled release of Doxorubicin under ideal physiological conditions. Another beneficial attribute of the complex is its synergistic potential in high drug loading and rapid burst of drug release under the influence of near-infrared (NIR) radiation (808 nm), thus proving a highly biocompatible thermo-chemotherapy for solid tumors (Mewada *et al.* (2013).

Cysteamine hydrochloride–protected C-dots have the potential to be used for drug delivery (Haloperidol) to treat neurodegerative diseases such as schizophrenia. Cysteamine hydrochloride (CysHCl) is used as a non-toxic linker to functionalize the C-dots before grafting Haloperidol onto the surface to facilitate the controlled release under physiological conditions. So far, a constant release of Haloperidol has been noted for more than 40 h. The C-dot–CysHCl– Haloperidol conjugate has shown much higher compatibility with MDCK cells at pH 7.2 in comparison to bare Haloperidol (Pandey *et al.* 2013).

Hollow luminescent carbon dots prepared from bovine serum albumin have also been tried for pH-controlled delivery of Doxorubicin by Wang *et al.* (2013).

3.2.4.4 Optoelectronics and In Vivo Biosensing Applications

The high resistance to photobleaching makes C-dots attractive materials for optoelectronics and in vivo biosensing applications (Abhilash, 2010). C-dots may also serve as a fluorescent sensor for the detection of metal ions and DNA, as emitting materials for light-emitting devices (LED). The interesting photoinduced electron transfer properties of C-dots should offer exciting opportunities for light energy conversion, photovoltaic devices and related applications.

3.2.4.5 Photocatalysis

Photocatalytic properties of C-dots can be used for the degradation of dye molecules. C-dots have peroxidase mimetic function and visible-light-sensitive photocatalytic activity for methyl orange degradation (Li 2010). In addition, a novel photocatalyst (TiO_2/C-dots) was obtained by combining C-dots with TiO_2 through an easy hydrothermal method. Remarkably, TiO_2/C-dots exhibit an excellent visible-light photocatalytic activity.

3.2.4.6 SERS

Au nanoparticles have been widely used as efficient enhancers for Raman scattering, and this phenomenon is known as the surface-enhanced Raman scattering (SERS) effect. Their enhancement effect is believed to be due to the excitation of the surface plasmon resonance (SPR) on the metal surface, which greatly strengthens the local electric field near the surface. Important applications of SERS can be found in chemistry, biology and material science as an ultra-sensitive detection technique capable of identifying trace molecules. Conventional SERS substrates are constructed by placing metal nanoparticles on planar surfaces, which thereby have a limited surface area. The replacement of these planar surfaces by spatially defined nanoporous surfaces with large surface areas will absorb more molecules for SERS signals and further improve the Raman detection sensitivity (Peng and Travas-Sejdic, 2009). C-dots are used as the basic building block for the controllable construction of macroscopic honeycomb structures via the assistance of a pre- assembled two-dimensional or three-dimensional silica colloid template. The resultant C-dot honeycombs can not only be used as a sacrificial template for the fabrication of new nanostructures made from other functional materials, but can also provide a new platform for the fabrication of a more efficient SERS substrate than the widely used Au nanofilm. (Raman enhancement effect can be dramatically improved by replacing the planar surface with a unique nanoporous super-aligned carbon nanotube film with cross stacking).

3.2.4.7 Health and Bio-Related Applications

With a new diversified concept of materials to be used in medical processes, medicine and medical implants, efforts are being directed towards use of new materials such as nano material, quantum dots, etc. in diagnostics, medical implants, biosensors, biochips and scaffolds for tissue engineering, sensing and bioimaging, etc. These applications need functional biomaterials that can be designed to produce a biological response, perform their desired function with respect to a medical therapy without eliciting any undesirable local or systemic effects in the recipient or beneficiary of that therapy, but generating the most appropriate beneficial cellular or tissue response in that specific situation. Optimizing the clinically relevant performance of that therapy can be termed as biocompatibility, the most attractive properties of C-dots. As carbon is the main constituent of almost all compounds, it is biocompatible, and due to its physicochemical and photochemical stability, it is widely preferred. C-dots are exerting a huge impact in both health- and bio-related applications because of their potential to serve as nontoxic replacements of traditional heavy metal–based quantum dots. Fluorescent C-dots have drawn considerable attention for biological labeling and biomedicine. Moreover, in pollutant detection, C-dots are also preferred over conventional inorganic quantum dots, because of their low toxicity, chemical inertness and biocompatibility.

3.3 Conclusions

Nano-sized graphene sheets and quantum carbon dots have been a major focus of research and development during the past decade involving their structural studies, properties, suitable green methodologies for fabrication, and above all, their various applications in almost all the fields of science. Since the discovery of nanographene and C-dots, many approaches have been performed to synthesize C-dots from top-down methods, such as arc-discharge, laser-ablation method, and electrochemical oxidation, to bottom-up methods, such as combustion, thermal decomposition, and microwave synthesis. Nevertheless, to obtain C-dots with high purity is a tough problem due to the complicated purification and separation procedures. Novel and unique properties, such as luminescence and semiconductor as well as low toxicity of these carbon configurations have been the major driving force for their studies. Although these graphene-derived materials hold great promise in healthcare and medicine, much work is still necessary to explore the full potential of these materials.

4

Identification and Characterization of Graphene

There is one thing even more vital to science than intelligent methods and that is, the sincere desire to find out the truth, whatever it may be.

Charles Pierce

4.1 Introduction

Chapter 2 included discussions on the specific unique properties and structure of graphene. The understanding of their structure, properties and modification to introduce a desired property, surface chemistry, etc., is possible only through the proper characterization. Thus, characterization presents many challenges some of which are not fully recognized, e.g., the need for real-time measurements in the environment of interest (*in situ*); the importance of surface and interface contamination, particle stability and time. Apart from analysis using equipment, complementary analysis tools and theory with expert applications of mathematical modelling is needed. In some cases, new concepts and new tools are needed to understand and characterize the fundamental nature of different types of graphene.

Hence, graphene's characterization is a necessary step in understanding some of its specific properties and impact of parameters responsible in the synthesis and its various applications. However, this newly developed, advanced material does not need the traditional characterization methods of a material, e.g., the assessment of molecular structure or composition, melting point, boiling point, vapor pressure, pH, solubility, water proportion and octanol, etc. Rather, characterization is needed by using various spectroscopic and microscopic methods. For the characterization of graphene, it is important to know the number of layers, the purity or presence of defects, structure, electronic characteristics, etc.

In this chapter, we look at the various analysis tools and their uses in identifying and characterizing graphene. The techniques used to understand the characterization's parameters of graphene are discussed in this chapter under five different sections as listed below.

A. MICROSCOPIC METHODS

Under this section, the following two major methods are considered:

i. SEM/STM/TEM to characterize the domain of growth and surface morphology
ii. AFM to know the surface morphology, thickness, uniformity of layers, domain growth

B. SPECTROSCOPIC METHODS

The spectroscopy technique is one of the important aspects of characterization, and encompasses many techniques. Under this section, the following techniques will be considered:

i. Raman spectra—to identify the graphene sheet and to get information about the number of layers
ii. FTIR—to assess the presence of functionalization groups
iii. UV-Vis Spectroscopy—to help in assessing the difference between graphene oxide and reduced graphene
iv. XRD (powder X-ray diffraction)—for analyzing the crystalline phase
v. XPS (X-ray Photoelectron Spectroscopy)— to analyze the surface chemical composition and bonding
vi. NMR—to obtainstructural information, e.g., sp^2 and sp^3 carbon information

vii. DLS (dynamic light scattering or photon correlation spectroscopy)—for size characterization or measuring the hydrodynamic size of molecules and submicron particles
viii. DPI (dual polarization interferometry)—to help in characterizing the surface interaction of graphene oxide with other molecules

C. OPTICAL ANALYSIS—for evaluating
i. Optical absorption and nonlinear Kerr effect
ii. Photoluminescence
iii. Optical band gap

D. MECHANICAL PROPERTIES MEASUREMENT
i. Young's Modulus
ii. Poisson's ratio
iii. Bulge test
iv. Surface tension (slack and self-tensioning in graphene membranes at $\Delta p = 0$ 102)
v. Gas leak rates in graphene membranes,
vi. Tunnelling of He atoms across a graphene sheet

E. THERMAL PROPERTIES and THERMAL EFFECT ANALYSIS
i. Thermal conductivity, and
ii. TGA and thermal stability

F. CHARACTERIZATION OF ELECTRICAL PROPERTIES
i. Electrical properties—conductivity, mobility, spin transport, etc.
ii. Electrochemical redox
iii. Carrier concentration: Anomalous Quantum Hall Effect
iv. Work function

We now look at the various analysis tools and their use in identifying and characterizing the graphene.

4.2 Microscopic Methods

4.2.1 SEM, STM and TEM Characterization of Graphene

For the characterization of graphene, it is necessary to study the surface morphology, pictorial views of arrangement of carbon atoms in the graphene, presence of grain boundaries to observed the domain size (which is important to understand the electrical properties of graphene sheet), layers of graphene (single layer, bilayer or multilayer). If graphene has been doped with some dopant, then its concentration as well as its position in the lattice should also be established. Such types of information can be obtained by studying the micrographs obtained with scanning electron microscope (SEM), scanning tunnelling microscope (STM), transmission electron microscope (TEM) and atomic force microscope (AFM).

Electron microscopes are scientific instruments that use a beam of energetic electrons to examine objects on a very fine scale. In the SEM, a focused beam of high-energy electrons are used instead of light to form an image. This examination can yield information about the topography (surface features of an object), morphology (shape and size of the particles making up the object), composition (the elements and compounds that the object is composed of and the relative amounts of them) and crystallographic information (how the atoms are arranged in the object). The SEM is also capable of performing analyses of selected point locations on the sample; this approach is especially useful in qualitatively or semi-quantitatively determining chemical compositions (using EDS), crystalline structure, and crystal orientations (using EBSD). The field emission scanning electron microscope (FE-SEM) images the sample surface by scanning it with a high-energy beam of electrons in a raster scan pattern.

Some SEM and TEM micrographs of graphene sheets are shown here to explain the advantages of these instruments. Figure 4.1 shows the graphene sheet grown on a copper substrate, which was pretreated. The SEM shows the effect on the morphology of graphene grown on copper, and which was cleaned by different processes. When copper was only thermally cleaned (A), the graphene growth is an irregular shape. With copper cleaned with nitric acid and then thermally treated (B), few hexagonal graphenes molecules grow on the copper. If the copper plate is electrochemically cleaned and then thermally treated (C), large regular hexagonal shapes of graphene grow on the copper substrate. This experiment suggests

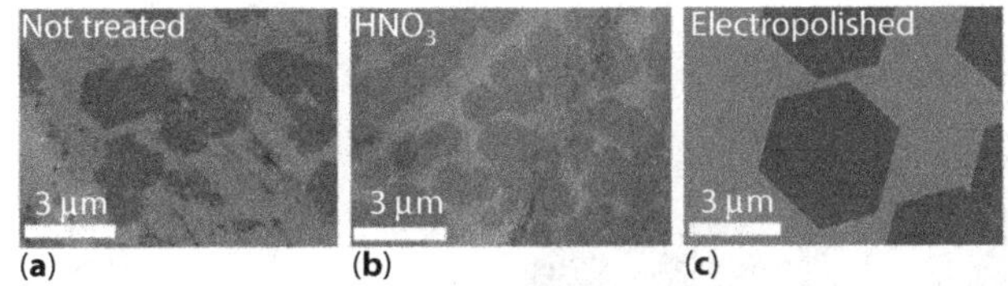

Figure 4.1 SEM micrographs showing the effect of copper substrate treatment (a) only thermal cleaning (b) treatment with nitric acid and (c) electrochemically cleaned on the growth of graphene film. Copyright by Elsevier Ltd., Courtesy: Vlassiouk *et al.*, Carbon 5 4, 5 8–6 7, 2013.

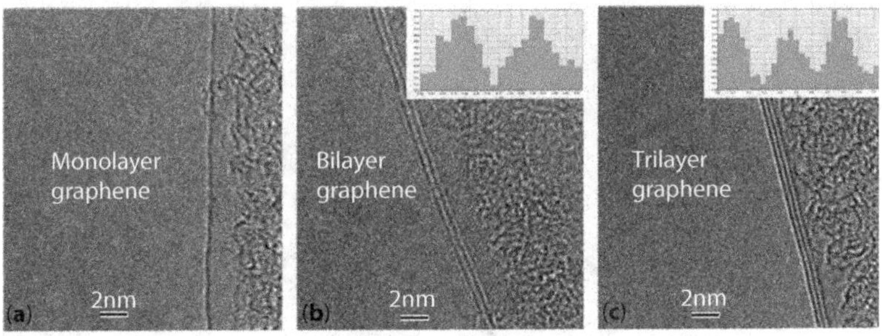

Figure 4.2 HRTEM images of graphene sheet at the folded edge showing (a) monolayer (b) bilayer and (c) trilayer graphene sheet synthesized from botanical precursor camphor. Copyright © 2011 Elsevier B.V. All rights reserved. Courtesy: Kalita *et al.*, *Physica E* 4, 2011, 1490–1493.

the impact of the morphology of copper substrate on the morphology of graphene molecules.

Unlike SEM, in transmission electron microscopy (TEM), a beam of electrons is transmitted through an ultra-thin specimen, interacting with the specimen as it passes through. TEMs are capable of imaging at a significantly higher resolution. Therefore, TEM forms a major analysis method for understanding the morphology of material. Nowadays, high-resolution TEM (HRTEM) are being used, where at high magnifications, the electron beam interactions with the atomic columns of a crystalline specimen lead to different contrast mechanism called phase contrast. This effect can be used for imaging the atomic distances in a properly oriented crystal. Being an additional technique to electron diffraction, it is a powerful method to gain further information such as crystal lattice distances and interface information to a limit below 1 nm.

Figure 4.3 STM of graphene flake (a) topography of three-layer flake (b) height profile showing separation of atomic layers (c) atomic resolution topography of decoupled graphene layer showing honeycomb structure and (d) Graphite. Courtesy Copyright © 2009, American Physical Society; Li *et al.*, *Physical Review Letters*, 2009, 102, 17804.

Information regarding the number of layers formed while synthesizing graphene sheets can be determined by studying the TEM. For example, Figure 4.2 shows the TEM micrographs of graphene sheets of different layers. The vertical single line denotes the number of layers. Figure 4.2a shows the presence of monolayer, while Figure 4.2b shows the presence of two layers, and Figure 4.2c shows growth of three-layer graphene.

An example of STM application to reveal the number of graphene layers and the honeycomb hexagonal structure of graphene can be visualised. How the graphitic carbon in a large number of layers can be distinguished from three layers of graphene sheets can also be visualised. By combining STM scanning tunnelling microscopy and spectroscopy together with Landau level spectroscopy, Lucian and Andrei (2008) were able to identify flakes of various thicknesses (Figure 4.3a) which reveals three layers separated in height by atomic steps of 0.35nm (Figure 4.3b). The decoupled flakes exhibit the distinctive honeycomb structure of graphene (Figure 4.3c), in clear contrast to the triangular structure seen on graphite (Figure 4.3d). In Figure 4.3c, a honeycomb hexagonal arrangement of six carbon atoms can also be seen.

4.2.2 AFM Characterization of Graphene

The suitability of the Atomic Force Microscope (AFM) is due to its high spatial resolution as well as various modes to probe different physical properties. AFM is capable of measuring nanoscale forces. A sharp tip at the end of a cantilever acts as a probe which is scanned line-by-line over a sample. The interaction between the surface of the sample and the sharp tip results in a force on the cantilever; the force is due to the van der Waals interaction, in which two uncharged atoms exert a force on each other when in close proximity. In contact mode, the tip is dragged across the surface, and the deflection is used to measure the height of the surface. In non-contact mode, the probe is driven at close to its resonant frequency and scanned across the surface. The force on the probe is used to determine the surface height (Figure 4.4).

AFM has been extensively used to study the surface morphology, thickness, uniformity of layer and domain growth of graphene.

The measurement of graphene by AFM can be done in three different modes:

i. **The conventional AC Mode**: This mode in combination with TERS (Tip Enhanced Raman Spectroscopy, an advanced optical method) helps in determining the exact number of stacked layers.

   AFM studies of graphene deposited onto mica surfaces have yielded some very interesting results in terms of surface topology (Lui *et al.* 2009).

   Figure 4.5 shows the AFM of graphene grown on nickel substrate. Figure 4.5A shows the grain size of crystalline graphene structure with average grain size of 936 nm and for Figure 4.5B, 564 nm. It is an example to show how the grain size of the film can be determined by the AFM technique.

ii. **Quantitative Imaging Mode**: A newly developed imaging mode (QI™) by JPK Instruments AG. The advantage of this mode is that it allows exact force control in both lateral and vertical direction, thus enabling sensitive imaging of loosely

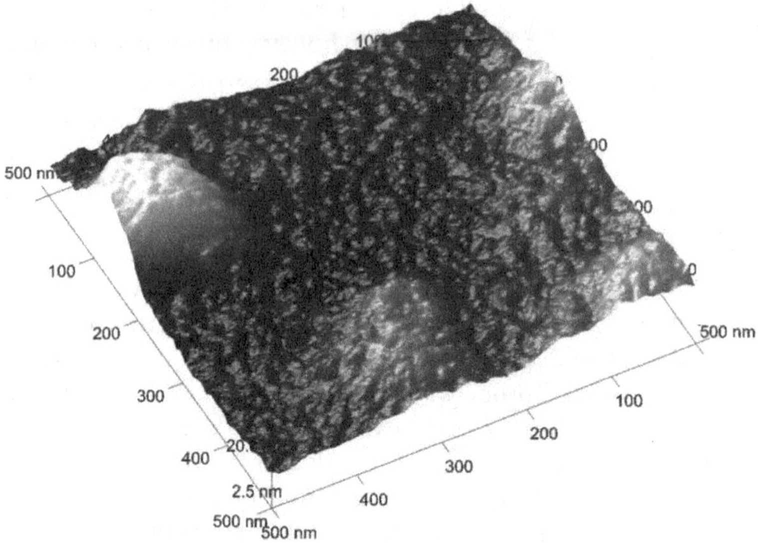

Figure 4.4 AFM image of graphene showing a topographical mapping of reduced GO on mica. The image shows that at some point agglomeration appears which creates an uneven structure with some spread peaks as evidenced by the bright fields in the image. (Courtesy: Joel Leffler, Towards graphene based transparent conductive coating; Master Thesis, Luleå University of Technology, 2013.)

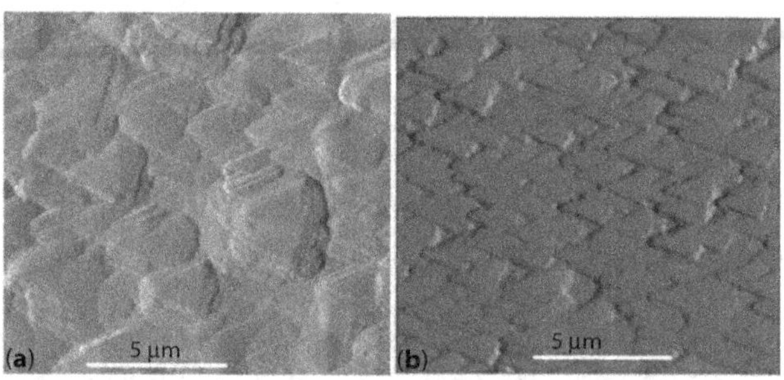

Figure 4.5 Typical characteristics of as-grown graphene (a) AFM image (amplitude contrast) of the sub/Ni-silicide/Graphene surface showing typical grain sizes in the range of 1–2 µm, (b) AFM image of the sample surface showing typical grain sizes of 0.5–1 µm. Copyright © 2013 The Authors. Published by Elsevier Ltd. Courtesy: Genhua Pan *et al.*, *Carbon* 65 (2013) 349–358.

attached graphene flakes. This helps in investigating several properties of graphene.

iii. **Contact Mode or Conducting AFM (CAFM):** Helps in electrical measurement. Unlike graphite that has high conducting properties due to delocalized electrons in between the layers, the single graphene layer shows an isolating behavior along the perpendicular direction. CAFM mode, in addition to topography data, also reveals the conductivity distribution.

It is worth mentioning here that to investigate the distinction between different numbers of graphene layers, a combination of AFM along with Raman spectroscopy or TERS (Tip Enhanced Raman Spectroscopy) is used. Moreover, many topographical properties can be investigated by scanning with Conductive AFM along with Kelvin Probe Microscopy, near-field scanning optical microscopy, scanning thermal microscopy, etc. The software for and supporting systems (NanoWizard®) are provided by JPK Instruments.

4.3 Spectroscopic Methods

4.3.1 Raman Spectroscopic Analysis of Graphene

Although TEM or STM can provide information about the number of layers present in the graphene sheet, sample preparation and the quality of the sample surface is very important to get accurate information about the presence of the number of layers present in the graphene sheet.

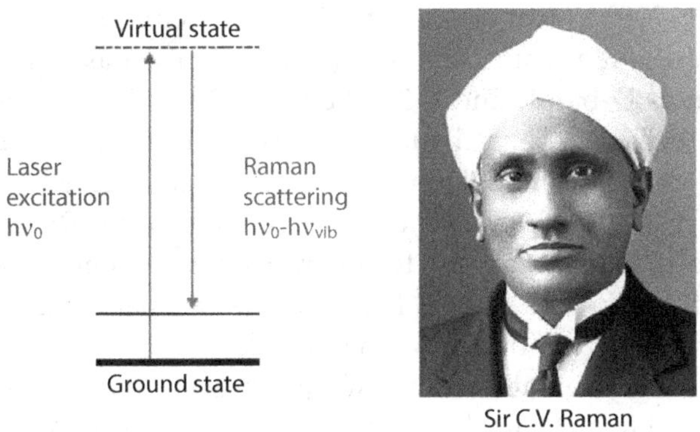

Figure 4.6 Schematic of excitation and de-excitation of electron due to Laser.

Raman spectroscopy is based on the observation made by Nobel Laureate C.V. Raman (1928), according to whom, "A small fraction of sunlight passing through a transparent medium is scattered in all directions, with a different wavelength than the incident radiation. These shifts depend upon the chemical structure of the molecules."

Using Raman spectroscopy, we can determine composition, crystallinity, lattice strain, defects and crystal size. There have been many advancements and improvisations in Raman spectroscopy, so as to suit characterization of different materials. Apart from conventional Surface Enhanced Raman Spectroscopy (SERS), in 2000 Tip Enhanced Raman Spectroscopy (TERS) was developed, which is similar to SERS but has much better lateral resolution (Stockle *et al.* 2000, Anderson 2000, Hayazawa *et al.* 2000 and Richards *et al.* 2003). In TERS, a laser illuminated metallic tip (analogous to the roughened metal surface used in SERS) is used that can enhance the intrinsically weak Raman signals by the generation of localized surface plasmon polaritons (SPPs). Moreover, TERS have been optimized and used as a tool to detect single molecules. TERS in combination with AFM has evolved as a device to characterize graphene.

Raman spectroscopy is a non-destructive and quick characterization technique to find out even the number of layers present in the graphene sheet. Raman spectroscopy can provide information such as whether the film is graphitic or diamond-like, and how many layers are present in the graphene film. Because of these advantages, Raman spectroscopy in the characterization of graphene has become very popular.

Raman scattering is one of the significant experimental technique to study the property of graphite/graphene. Malard *et al.* (2009) have given a good account of the Raman spectroscopy of graphene and theory to explain the various bands observed with graphene.

The first report of Raman spectra of graphene was given by Tuinstra and Koenig (1970) as can be seen in Figure 4.7. Graphite shows the three most intense Raman features at ~1582 cm^{-1} (known as G band), ~ 1350 cm^{-1} (known as D-band) and ~ 2700 cm^{-1} (2D band, which is also called as G^1 band) because the G^1 band is an overtone of the D band. A weak band at ~3248 cm^{-1}, called $2D^1$ band, which is also referred as G^{11} band, is an overtone of D^1(1620 cm^{-1}) mode. The second-order Raman spectrum (G^1or 2D, i.e., 2700 cm^{-1}) is due to the overtones and combinations of the disorder-induced bands. This is from double-resonant Raman scattering by two phonons instead of a phonon and a defect.

Since 2D band originates from a two-phonon double resonance process, it is interrelated to the band structure of graphene layers. D band consists

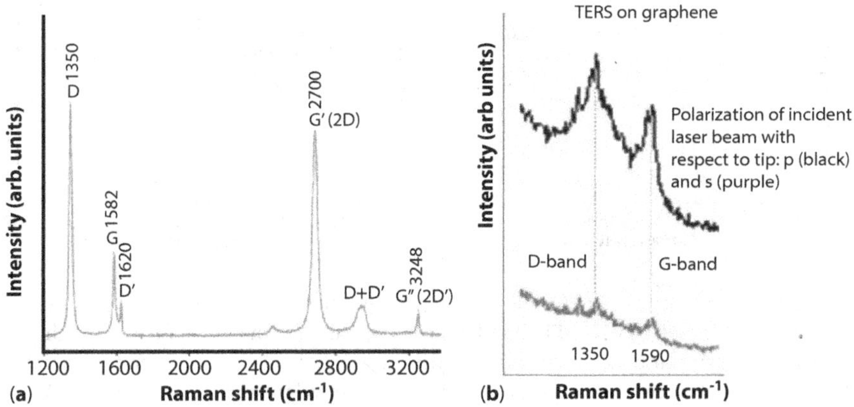

Figure 4.7 (a) Raman spectrum of graphene (Rights managed by AIP Publishing LLC. *Courtesy: Tuinstra and Koenig. J. Phys. Chem.* ***53****, 1126, 1970*) and (b) TERS spectra of graphene obtained using a gold tip and tuning fork AFM feedback when polarized incident light is directed along the tip axis. Copyright © 2014 American Physical Society, Courtesy: Gerber *et al.*, *Phys. Rev. Lett.* 113, 050401 (2014).

of one-elastic and one-inelastic scatterings, while the G^1 band consists of two inelastic scatterings.

The G-band is due to doubly degenerate E_{2g} mode at the Brillouin zone centre, whereas D-band arises from defect mediated zone-edge (near K-point) phonons. D-band is also considered due to the presence of sp^3 carbon, because pure diamond, which contains 100% sp^3, shows D-band around 1350 cm^{-1}. Similarly, G-band (around 1582 cm^{-1}) is considered due to the presence of sp^2 carbon. Thus, the percentage of sp^3 in any carbon nanotubes or graphene can be evaluated by comparing the area of the peak height of D-band (obtained with carbon materials) with D-band intensity obtained with pure diamond. Likewise, the concentration of sp^2 in graphene can be calculated by comparing the area of G-band obtained with HOPG (which contains 100% sp^2 carbon). It is interesting that intensity of the 2D band can be finite even if the D-mode intensity is negligible (Figure 4.8A).

It is also worth perceiving that Raman spectroscopy does not depend on the substrate used. This is because the Raman spectrum is an intrinsic characteristic of graphene, and Raman intensity is proportional to the thickness of the graphene sheet and the laser of a 532 nm wavelength can penetrate up to 50 nm of graphene sheet, which is approximately 150 layers. Wang *et al.* (2008) have reported the Raman spectrum of graphene

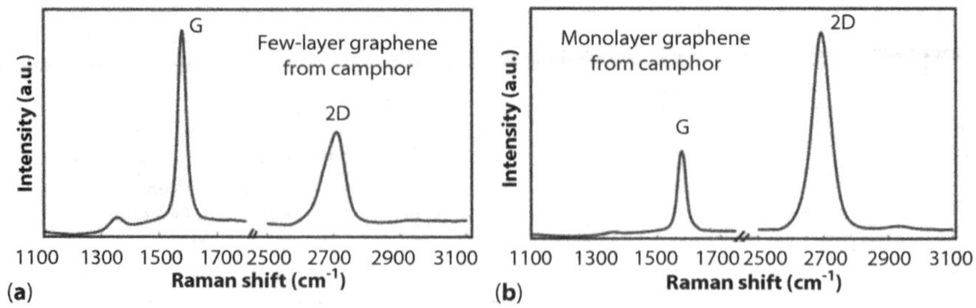

Figure 4.8 Raman spectra of (a) multilayer layer graphene sheet showing presence of 2D band as well as negligibly small D-band at 1350 cm^{-1} and (b) monolayer graphene sheet synthesized from camphor on copper foil. The intensity of 2D band is much bigger than that of G, which is reverse for multilayer graphene. Copyright © 2011 Elsevier B.V. All rights reserved. Courtesy Kalita *et al. Physica E* 43 (2011) 1490–1493.

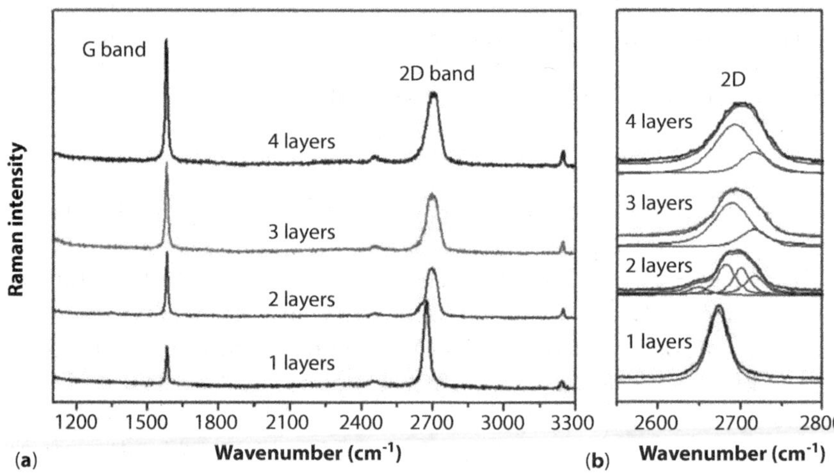

Figure 4.9 (a) Raman spectra of graphene with 1, 2, 3 and 4 layers. (b) The enlarged 2D band regions with curve fitting *Topics Appl. Physics* 111, 673–709 (2008). Copyright © 2007, Springer-Verlag Berlin Heidelberg, Courtesy: Charlier *et al.* in edited book: A. Jorio, G. Dresselhaus, M. S. Dresselhaus (Eds.): *Carbon Nanotubes Topics Appl. Physics* 111, 673–709 (2008).

sheet with different layers (Figure 4.9). The sharp G peak in comparison to 2D peak suggests a high crystalline graphitic structure.

The advantage of this 2D peak is that its shape, position and relative peak height with respect to G-band can distinguish between single and multilayer samples. From Figure 4.9, peak heights of 2D band, G-band and shift in spectrum at 2700 cm^{-1} for various layers of graphene were calculated and the results are plotted in Figure 4.10. This was done to establish a relationship between these peak heights and the number of layers.

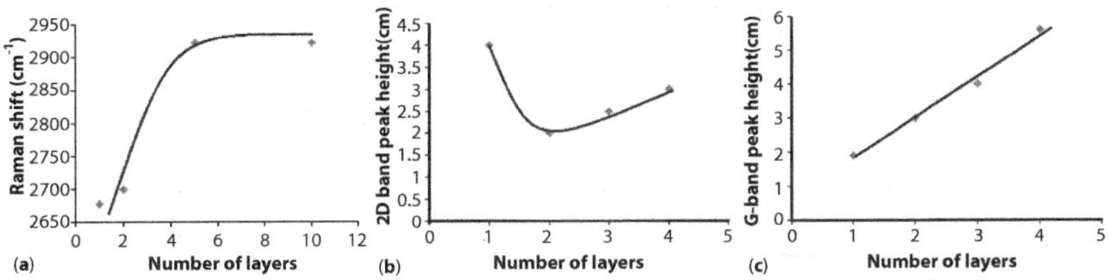

Figure 4.10 Graph is drawn from data presented in Figure 4.9. (a) Raman shift (cm^{-1}) around 2700 cm^{-1} obtained with different layers of graphene sheet (b) Likewise plot of 2D peak height (cm) obtained with different layers of graphene versus number of layers and (c) Plot of G-band peak heights (cm) obtained with graphene of different layers.

Raman shift at around 2700 cm^{-1} on plotting for different numbers of graphene layers (Figure 4.10A) shows that as the number of layers increases, the 2D-band moves lower than 2700cm^{-1} and after about five layers, it remains constant at 2700cm^{-1}. If the height of the 2D peak is measured for various layers of the graphene sheet, it shows a minimum for bilayer graphene (Figure 4.10B). Another interesting observation is obtained with G-band peak heights versus the number of layers (Figure 4.10C), which shows a linear increase in height with number of layers. Moreover, depending upon the intensity of the second-order Raman spectrum of graphene, it can be observed that 2D peak is much more intense than that of the G peak in the case of a single graphene layer, and is reduced with an increase in the number of graphene layers (Figure 4.9). Qualitatively it is observed that two to three layers on a graphene sheet shows the ratio of G-peak to 2D peak < 1.5 and those >1.5 shows multilayer graphene sheet, e.g., 10–15 layers. These observations suggest that Raman spectrum of G -band and 2D band can provide information about the number of layers present in the graphene sheet, without damaging the film.

Wang *et al.* (2009) have shown that even after chemical functionalization of graphene sheet to make them hydrophilic or organophilic, the characteristics of Raman spectra (Figure 4.11) did not alter. They exhibited a strong D line around 1350 cm^{-1}, representing a breathing mode of j-point photons of A_{1g} symmetry, and a relatively weak G line around 1582 cm^{-1}, which should be assigned to the first order scattering of the E_{2g} phonon of sp^2 C atoms.

4.3.2 FTIR Analysis of Graphene

Unlike UV-Vis Spectroscopy, where a monochromatic light beam is used for absorption studies, in Fourier transform spectroscopy, a beam containing

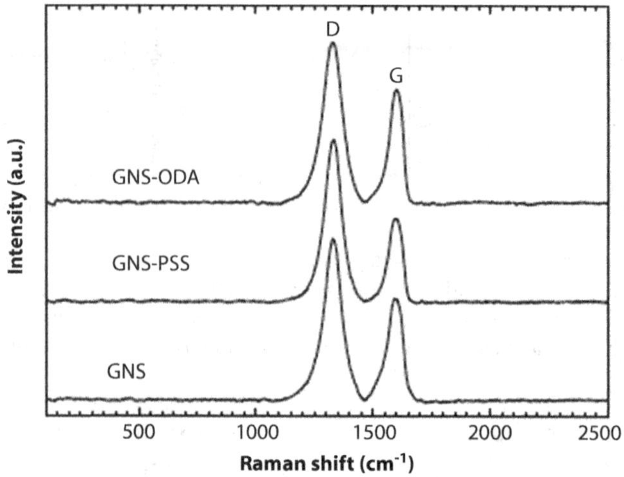

Figure 4.11 Raman spectra of the pristine graphene nanosheets (GNS), hydrophilic graphene nano sheets (GNS-PSS) and hydrophobic graphene nano sheets (GNS-ODA). Courtesy: Wang *et al.*, *Carbon*, 47 (2009), 1359–1364.

many frequencies of light at once is used and the beam absorbed by the sample is measured and modified to contain a different combination of frequencies, giving a second data point; this process is repeated many times and finally a computer does the backwards calculations of these data to infer what absorption is at each wavelength. The resulting spectrum represents the molecular absorption and transmission of the sample creates a molecular fingerprint. It is used to identify the presence of certain functional groups present in a molecule. Moreover, this technique is a unique collection of absorption bands that helps in confirming the identity of a pure compound or to detect the presence of specific impurities. In the case of graphene analysis, it helps in characterizing the graphite, graphene oxide, reduced graphene and functionalized graphene.

FTIR spectra of graphite does not exhibit a significant peak, whereas the presence of different types of oxygen functionalities in graphene oxide can be confirmed by (O-H stretching vibrations) at 3400 cm^{-1} and (stretching vibrations from C=O) at 1720 cm^{-1} and at 1600 cm^{-1} (skeletal vibrations from unoxidized graphitic domains), at 1220 cm^{-1} (C-OH stretching vibrations), and at 1060 cm^{-1} (C-O stretching vibrations) (Xu *et al.* 2008).

The FTIR peak of reduced graphene exhibits the O-H stretching vibrations at 3400 cm^{-1} which gets significantly reduced due to de-oxygenation. But stretching vibrations from C=O at 1720 cm^{-1} remains the same

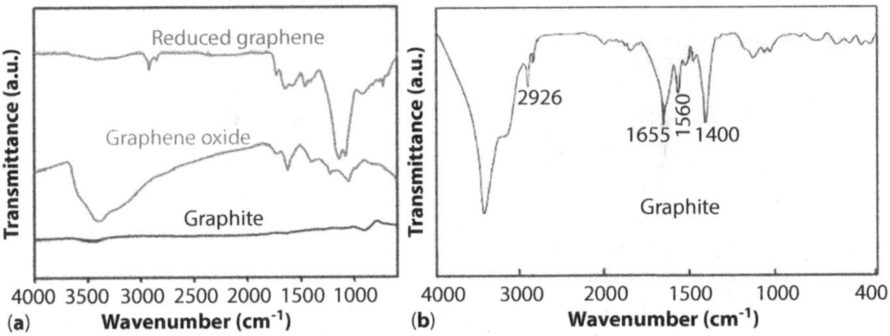

Figure 4.12 (a) FTIR spectra of graphite, graphene oxide and reduced graphene and (b) FTIR spectra of pure graphite. (Copyright © 2010, Springer Science+Business Media, LLC; Courtesy: Konwer *et al.*, *Journal of Materials Science Materials in Electronics*, 2011; 22(8):1154–116)

and C-O stretching vibrations at 1060 cm^{-1} becomes sharper, because of remaining carboxyl groups even after hydrazine reduction (Li *et al.*, 2008). A typical FTIR spectra of graphite, graphene oxide and reduced graphene is presented in Figure 4.12.

4.3.3 UV-Vis Spectroscopic Analysis of Graphene

Under the UV-Vis spectrophotometer, lights of a wavelength of 200nm–1100 nm are used to probe electronic transitions. Spectrophotometer with wavelengths lower than 200 nm is not observed because oxygen starts absorbing light at wavelength below 200 nm. UV-Vis spectroscopy has been very useful in characterizing the graphene oxide (GO) and reduced graphene.

The attachment of oxygen to the graphene layers increases the polarity of the layers, which in turn increases their solubility in water. This results in a change of color, which depends on the concentration of GO in the solution. Usually it is shades of a brownish color of different intensities, having an absorption maximum at ca 230 nm for a well oxidized material (Baladin *et al.* 2007).

UV-Vis spectra of reduced graphene at ca 230 nm shifts to a higher wavelength. This is due to the oxygen attached on the layers being desorbed by the reducing agent. As an effect of this, the solution of GO change color from brown to black. When graphene is reduced, the absorption increases at all wavelengths as the reduction progresses. The high absorbance at all wavelengths is in accordance with the solution becoming black, i.e., the solution absorbs light at all wavelengths.

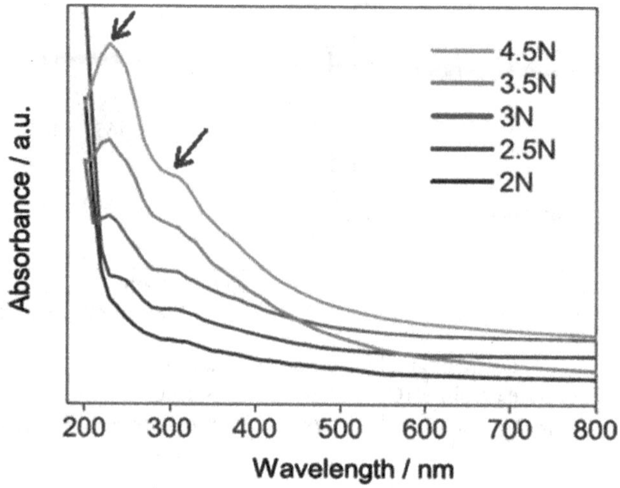

Figure 4.13 UV-Vis spectra of aqueous solution of GO showing two characteristic features to identify GO (i) a shoulder at ~ 310 nm corresponding to n–π* plasmon peak and the other (ii) at 230 nm and corresponds to π–π* plasmon peak. Rights managed by AIP Publishing LLC. Courtesy Qi Lai *et al. AIP Advances* 2, 032146 (2012); http://dx.doi.org/10.1063/1.4747817.

UV-Vis spectra has been used to find out whether graphene can be oxidized by different concentrations of $KMnO_4$, and it was observed that a 4.5N $KMnO_4$ solution was most effective in oxidizing graphene as shown by a sharp peak at 230 nm and a shoulder at 310 nm. The peak at 230 nm was due to π–π* plasmon interaction, whereas a shoulder at 310 nm is due to π–π* plasmons interaction. By examining the graphene films the oxidation process could be related to the number of layers of graphene. Wu *et al.* (2008) concluded that one- to three-layer GO shows a single peak at 230 nm, while multilayer GO of 4–10 layers or more exhibits a shoulder at 310 nm. As the number of layers increases, the intensity of the shoulder of multilayer GO tends to decrease (Figure 4.13).

4.3.4 XRD Analysis of Graphene

XRD is a very important experimental technique that has long been used to address all issues related to the crystal structure of solids, including lattice constants and geometry, identification of unknown materials, orientation of single crystals, preferred orientation of polycrystals, defects, stresses, etc. When X-rays interact with a crystalline substance, the result is a diffraction pattern. Hence, X-ray diffraction analysis is done to know the structure of the pristine graphene.

Graphene is an isolated atomic plane of graphite. From this perspective, graphene has been known since the invention of X-ray crystallography. There are two major peaks exhibited in X-ray diffraction analysis of graphene-related materials; graphite shows a characteristic signature peak at 27 degrees, whereas the peak for GO is around 10 degrees corresponding to (001) diffraction peak of GO. The shift of the diffraction peak of graphite from 27 degree to 10 degree for GO means that the spacing between the graphene layers has increased; oxidation is supposed to create spaces in between the layers, and a large distance is indicated by diffraction peaks at low angles. Moreover, the peak around 10 degrees is absent when GO is reduced, and the structure there after depends on which substance is used as the reducing agent. For example, Lai *et al.* (2012) have shown the usefulness of XRD analysis in characterizing the crystal structure of chemically functionalized graphene sheets. As can be seen in Figure 4.14, the pristine graphene nanosheets show a (002) diffraction peak at 26° (2θ), corresponding to a d-spacing of 0.34 nm, which is almost the same as that of graphite. Both hydrophilic graphene nanosheets (GNS-PSS)).PSS- [poly(sodium 4-styrensulfonate)] and ODA- (Octadecylamine) functionalized organophilic graphene nanosheets (GNS-ODA) show a (002) diffraction peak at 21.5°, corresponding to a d-spacing of 0.41 nm, which is larger than that of the pristine graphene nanosheets. The scientists proposed that the surface grafting with ODA molecules proceed on the surface of the basal planes through epoxy ring-opening, as earlier suggested by Lai *et al.* (2012)

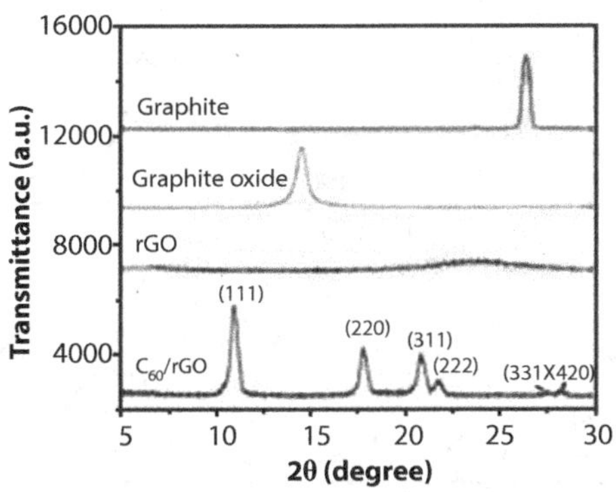

Figure 4.14 XRD patterns of graphene nanosheet (GNS) pristine graphene nanosheets reduced by hydroquinone; (GNS-PSS) hydrophilic graphene nanosheets reduced by hydrazine hydrate and (GNS-ODA) organophilic graphene nanosheets functionalized with Octadecylamine and reduced by hydroquinone. Copyright © 2013, Rights Managed by Nature Publishing Group, Courtesy: Zhang *et al.*, *Scientific Report*, 2013, 3.

4.3.5 XPS of Graphene

X-ray photoelectron spectroscopy (XPS) is a surface-sensitive technique that is used to analyze the surface chemical composition, depth profile of the composition and bonding.

Chen *et al.* (2011), while studying the ability of graphene films grown by chemical vapor deposition to protect the surface of the metallic growth substrates of Cu and Cu/Ni alloy from air oxidation, have shown by XPS studies that the metal surface is protected from oxidation even after heating at 200°C in air for up to 4 h. Figure 4.15A shows XPS spectra of copper plate over which graphene is deposited. These spectra show that copper has not oxidized after thermal treatment at 200°C for 4 h. Figure 4.15B shows the copper plate when thermally treated at 200°C for 4 h, in which it forms various oxides of copper (Cu_2O, CuO, $Cu(OH)_2$). These peaks are obtained in addition to peak corresponding for the pure copper, i.e., $Cu(2p_{3/2})$ and $Cu(2p_{1/2})$. This experiment suggests that a graphene layer can protect metal from oxidation.

Another application of XPS was demonstrated by Yang *et al.* (2009) while measuring the chemical analysis of graphene oxide films after heat and chemical treatments. They could exhibit significant reduction of the oxygen content of the graphene films; heating in ultra-high vacuum was found to be particularly effective (Figure 4.16). Based on XPS data, they

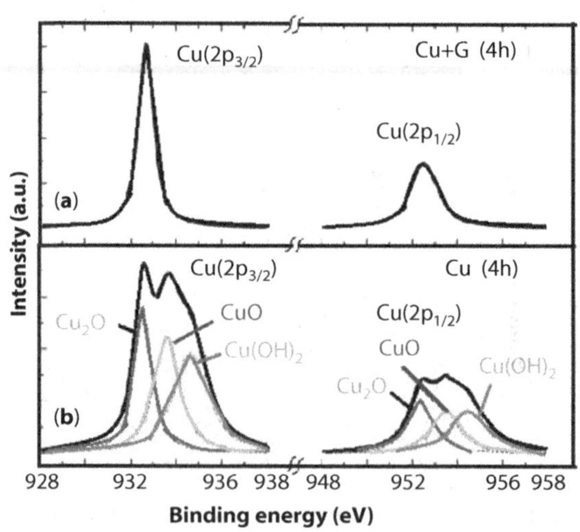

Figure 4.15 XPS core-level Cu2p spectrum of coated (upper) and uncoated (lower) Cu foil after air anneal at 200°C for 4 h. Copyright 2012, American Chemical Society, Courtesy: Chen *et al.*, *ACS nano* 2011 Vol: 5(2):1321. DOI: 10.1021/nn103028d.

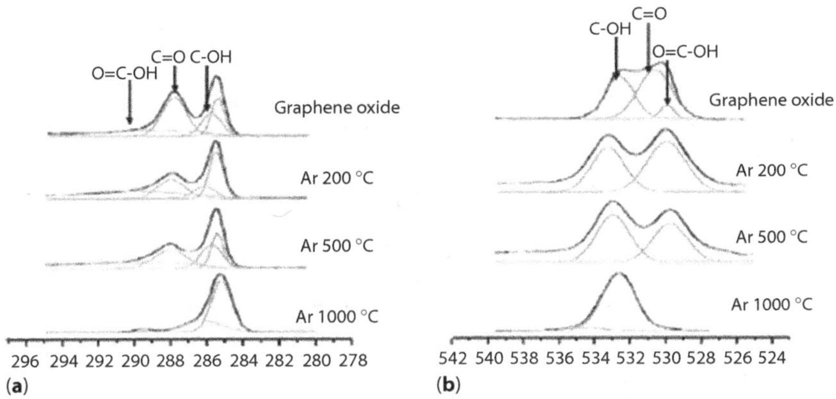

Figure 4.16 XPS spectra of graphene oxide films on Si substrate after 200, 500 and 1000°C treatment in flowing Argon: (a) comparison of C1s peaks and (b) composition of 01s peaks. Courtesy: Yang *et al.*, *Carbon* 47 (2009) 145–152.

concluded that there was a significant reduction of such films, with the maximum reduction at high temperature and heating under UHV being particularly advantageous. XPS spectra was key in assigning the degree of reduction.

4.3.6 NMR Analysis of Graphene

NMR analysis is undertaken to acheive structural information, e.g., sp^2 and sp^3 carbon information.

Graphene oxide (GO) shows amorphous and non-stoichiometric nature. This property makes it difficult to understand its actual structure. In addition, GO does not show the same level of oxidation. One typical NMR spectra of graphene oxide nano sheet (GONS) is shown in Figure 4.17.

Nevertheless, a structure proposed by Lerf *et al.* (1997, 1998) has been accepted. This structure is based on the use of solid-state nuclear magnetic resonance (NMR) spectroscopy. Lerf and colleagues proposed that hydroxyl group (-OH), epoxy ($-O^-$) is spread over the planes of graphene. Carboxyl group and keto group exist mainly at the edges of the graphene plane. There have been questions as to whether the sp^2 group clusters in aromatic form. Due to various developments in the instrumentation and magnetic power, NMR has become a very useful tool to understand the structure of graphene. In fact, Lerf's model is based on the ^{13}C NMR results. Three major peaks (not shown here) have been assigned to epoxy group (60ppm), hydroxyl group (70ppm) and sp^2 carbon atom (129–130ppm).

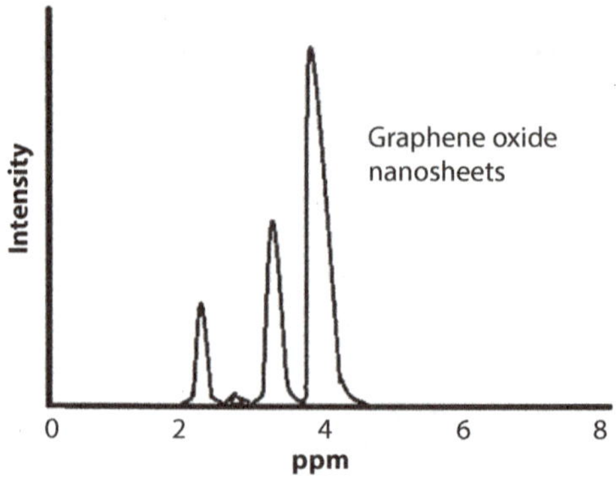

Figure 4.17 ^{1}H-NMR spectra of graphene oxide.

In NMR studies, it was noted that full-width-at-half-maximum height of the water peak remains nearly constant in the ^{1}H NMR spectrum (approx. 2.8 kHz) across a wide temperature (123–473 K), indicating very strong interactions between the water and the GO. This is likely a key factor in maintaining the stacked structure of GO.

^{13}C NMR of functionalized GO shows about 60% of the carbon atoms are sp^3 or aromatic carbons. It also contains carbonyl groups in lactols. Carbon are hybridized and oxidized, mostly in the form of alcohols and epoxides. Some are also as lactols. The remaining 40% are alkene atoms (sp^2-hybridized). Others are mostly as unfunctionalized esters, acids, and ketones.

4.3.7 DLS of Graphene

DLS (dynamic light scattering or photon correlation spectroscopy) is used to measure the hydrodynamic size (from <1 nm to 1 um) and size distributions of macromolecules and nanoparticles need to be measured quickly and easily. DLS measures variation in scattered intensity with time at a fixed scattering angle (typically 90°), while static light scattering measures scattered intensity as a function of angle. Colloidal GO is usually illuminated by a He-Ne laser at $\lambda = 633$nm and the scattered light is collected at 173° to the laser beam. By this method, Natalya *et al.* (2014) measured the average hydrodynamic size of d_{HD} with an accuracy of 4–5%.

4.3.8 DPI of Graphene

DPI (dual polarisation interferometry) is a technique that helps in characterizing at real time the surface interaction of graphene oxide with other molecules. Frost *et al.* (2012) have done real-time monitoring of interaction between graphene oxide monolayers and lipid membranes using this method.

DPI works by focusing laser light into two waveguides, (i) a "sensing" waveguide with an exposed surface and (ii) a second waveguide that functions to maintain a reference beam; thus forming two-dimensional interference pattern. The DPI technique rotates the polarization of the laser, to alternately excite two polarization modes of the waveguides. Measurement of the interferogram for both polarizations allows both the refractive index and the thickness of the adsorbed layer to be calculated. The polarization can be switched on rapidly, allowing real-time measurements of chemical reactions taking place. The technique is quantitative and real-time (10 Hz) with a dimensional resolution of 0.01 nm. DPI technique has become very handy to understand the mechanism by which different types of proteins interacts with graphene and graphene oxide. This information is useful in making biosensors using graphene sheet.

4.4 Optical Property Analysis

4.4.1 Optical Absorption and Nonlinear Kerr effect

Nonlinear optical effects arise when electronic motion in a strong electromagnetic field is non-harmonic. A material under the influence of electric field if illuminated results in refractive index changes. This effect was observed by John Kerr in 1875, and hence this is referred to as a Kerr effect or quadratic electro-optic effect. The refractive index changes linearly with the applied electric field. There are two types of Kerr effect: Kerr electro-optic effect under DC electric field or optical Kerr effect under AC electric field. Under the influence of DC electric field, the material becomes birefringent, showing different indices of refraction of light polarized parallel or perpendicular to the applied field. The difference in refractive index Δn is given by the following equation:

$$\Delta n = \lambda K E^2$$

Where λ is wavelength of the light, K is Kerr constant and E is electric field. The difference in the refractive index causes the material to behave like a so-called wave-plate, when the light falls on the material in direction perpendicular to the electric field. The importance of the Kerr effect is that when the material is between two crossed (i.e., perpendicular to each other) linear polarizers, material will not be able to transmit light if the electric field is switched off. By adjusting the magnitude of the electric field, all light will be transmitted. In other words, transmission of light through the material can be tuned by adjusting the magnitude of the electric field. It is interesting to know that in AC Kerr effect, a strong laser light falling on the material can create a modulating electric field without applying an electric field.

Multilayer graphene shows an increase in transmittance due to saturable absorption as well as nonlinear phase shift. It has been shown that graphene possesses a nonlinear refractive index of $10^{-7}cm^2\ W^{-1}$, which is nine orders of magnitude larger than bulk dielectrics. These results support the application of graphene-based nonlinear photonics. In addition, it has been reported by Levallois *et al.* (2011) that polarized light, when it passes through a monolayer graphene sheet under a magnetic field, is rotated by almost 6° to its initial angle (Figure 4.18). It is surprising that such behavior is observed only with thicker material. This behavior of graphene can be useful for switching even infrared light under electrical and magnetic fields. It is surprising that graphene has a bigger Faraday rotation per atomic layer than any other material by a factor of 10. Because the Faraday angle is proportional to the thickness of the material, in graphene—being just one atomic layer thick—such behavior was unexpected. Faraday rotation was measured by passing infrared light through a polarizing filter to create a linearly polarized beam, which was allowed to pass through a graphene sample with a magnetic field perpendicular to its surface. The transmitted light was allowed to pass through a second polarizing filter. Because the polarizations of the two filters were 90° apart, no light was observed. As the polarization of the light rotated as it passed through the graphene, the

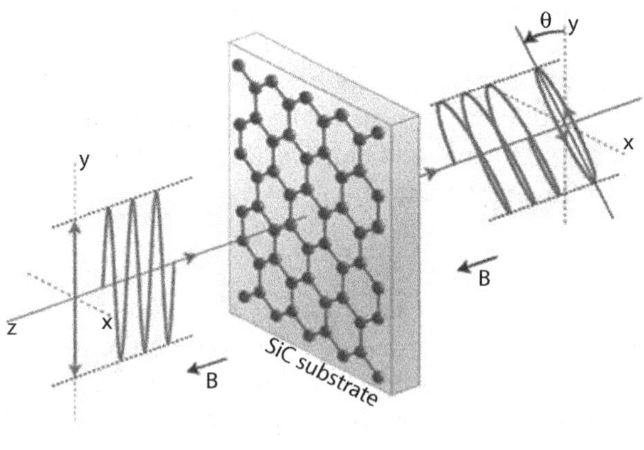

Figure 4.18 Showing how polarized light when passes through the graphene the transmitted light shifts angle by 6°. Copyright © 2013, Rights Managed by Nature Publishing Group, Courtesy: Crassee *et al.*, *Nature Physics* 7,48–51,2011.

angle at which no light detected was shifted by the Faraday angle. Graphene thus can be used as a switch to allow light to pass in one direction and not in the opposite direction. Such types of optical diodes (Faraday isolators), are not available for infrared light. The other advantage of such type of magneto-optical switch is that by altering the magnetic field direction, the Faraday rotation can be reversed. This is possible with graphene due to its ability to change the sign of its charge carriers from negative to positive by simply applying an electric field (due to its ambipolar electric field effect).

The Faraday effect and the magneto-optical Kerr effect are needed in optical communications, data storage and computing. In addition, these devices are expensive and large in sizes. But it is expected some developments to graphene could be made to overcome its inherent problems, after which this material could be of immense benefit for such switching devices. For example, before graphene could be used for switching purposes, one may need at least 10 layers of graphene to achieve a rotation of 45°. In addition, since graphene absorbs infrared light, there is a possibility of losing intensity of signal in the device.

4.4.2 Photoluminescence/Blue-Photoluminescence

Graphene oxide is an intermediate on the route to chemically derived graphene and exhibits luminescence in the visible (Vis) and near-infrared (NIR) wavelengths range. This property is of importance for many (bio) analytical applications (because it is easily dispersible in water), biosensing and fluorescence tags, and development of low-cost opto-electronic devices. Blue fluorescence is especially useful for display and lighting applications. This opto-electronic property is due to the presence of sp^2 clusters. The PL in such carbon systems is due to the recombination of localized e–h pairs in sp^2 clusters. These centers behave as the luminescence centers or chromophores. Photoluminescence (PL) emission is achieved by controlling the nature of sp^2 sites. Blue PL is due to the radiative recombination of e–h pairs generated within localized states.

In order to confirm that a sp^2 cluster is responsible to show intense PL, Goki Eda *et al.* (2010) studied the PL of graphene oxide (GO) and the effect of its gradual reduction of graphene oxide to graphene (rGO). They observed that pure GO shows the very small PL spectra and GO reduced by hydrazine shows a gradual increase in PL spectra with time of reduction of GO to graphene (rGO). The highest PL spectra was obtained with GO reduced for 60 min (Figure 4.19). A reduction of GO will increase the sp^2 clusters in graphene (rGO) and hence shows an increase in PL intensity. Interestingly, the changes in fluorescence intensity with hydrazine does not

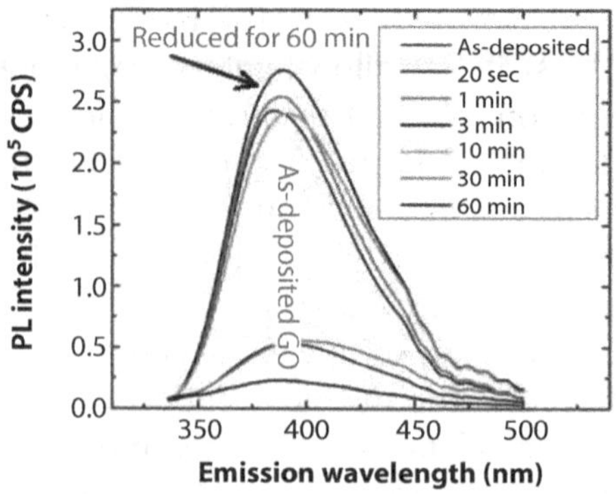

Figure 4.19 Blue photoluminescence from chemically derived graphene oxide. Copyright © 2010 Wiley-VCH Verlag GmbH and Co. KGaA, Weinheim, Courtesy: Goki Eda *et al.*, *Adv. Mater.* 2010, 22, 505–509.

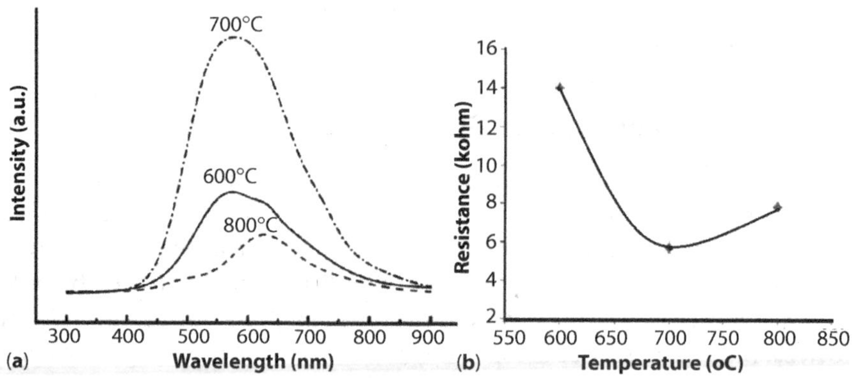

Figure 4.20 (a) Room temperature PL spectra of graphene thin films prepared by thermal reduction of GO at different temperatures. (b) A graph of resistance versus temperature drawn from data provided in the same paper. Copyright © 2010 Elsevier B.V. All rights reserved. Courtesy: Cuong *et al.*, *Materials Letters* 64, 765–767, 2010.

depend on the excitation/emission wavelength. Such property of GO/rGO may be useful for designing a probe compatible with optical equipment (light sources, filters, etc.)

It has been suggested by Tran Viet Cuong *et al.* that photoluminescence also depends on the temperature at which GO has been reduced (Figure 4.20). It is interesting to note that these PL spectra are similar to spectra observed from amorphous carbon. Furthermore, the plot of resistance of graphene film versus temperature (Figure 4.20B) shows that lower resistance gives the highest PL peak at 700°C. SEM of the graphene film

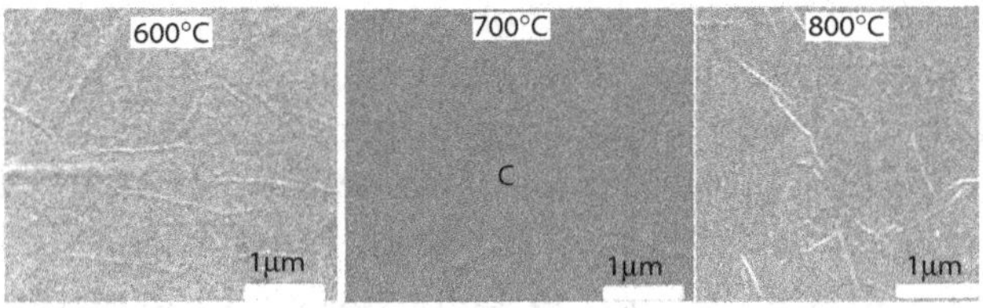

Figure 4.21 SEM micrographs of graphene films prepared at 600°C, 700°C and 800°C. Copyright © 2010 Elsevier B.V. All rights reserved. Courtesy: Cuong *et al.*, *Materials Letters* 64, 765–767, 2010.

(Figure 4.21) shows that film prepared at 700°C is more uniform, with almost negligible wrinkles as compared to other films prepared 600°C or 800°C (Figure 4.21). Though Tran Viet Cuong *et al.* (2010) explained this behavior on the basis of sp^2 and sp^3 carbon present in the film, after examining the SEM micrographs and the resistance of the film, it would perhaps be better explained on the basis of uniformity of the graphene films. It is expected that graphene film with lower amounts of grain boundaries or larger domain sizes should show the least resistance. It would be appropriate therefore to explain the PL spectra on the basis of grain boundaries or domain size. This is because smaller grain boundaries or larger domain size facilitates recombination of photoexcited electron-hole pairs, giving a better PL spectra. Temperature alone, as suggested by Tran Viet Cuong *et al.*, may not be intrinsically responsible for showing higher PL spectra.

Shang *et al.* (2012) did time-resolved fluorescence measurements of graphene oxide in water that showed multi-exponential decay kinetics ranging from 1 ps to 2 ns. They suggested that the origin of fluorescence is due to recombination of electron/hole created due to the excitation of electron from the upper level of valence band (i.e., creation of holes in the valance band) to the lower level of conduction band.

4.4.3 Optical Band Gap

Graphene seems to demonstrate metallic as well as semiconducting properties. Though some concepts based on a theoretical analysis suggest that graphene has a zero band gap and can be varied to about 0.3eV, there is still a need to look into the band formation in monolayer graphene and multilayer graphene. As per the conventional concept, band gap originates from the interaction of orbital levels of the electrons. The greater the interaction, the larger the band gap of the material. Assuming this concept to be applicable to graphene, the band picture present in graphene becomes confusing. In monolayer graphene, electrons involved with the σ-bond are not accepted to contribute towards the formation of conduction and valence band. On the other hand, electrons of π-orbital of two neighboring carbon atoms can interact to produce two energy levels, i.e., conduction and valence band. Considering this picture of carbon atoms, in a monolayer of graphene, π–π interactions can take place within the hexagonal structure of one ring (Figure 4.22X) or with carbon of the adjacent ring (not shown in figure). However, the distance between the two π-π electrons irrespective of being within one hexagonal ring or an adjacent hexagonal ring, may be almost of the same magnitude. Considering this picture of monolayer graphene, one would expect a very large band gap, because the distance of interactions between two π-electrons is very small (0.142 nm). This conclusion may appear to be contradictory because monolayer graphene is known to be highly conducting. The author of this book has observed that carbon nanotube shows two types of band gaps: direct band gap (which can be as high as 2eV) and indirect band gap (which can be as low as zero). If this is correct then high conductivity of graphene must be due to the presence of indirect band gap.

If this concept is applied with bilayer graphene, in addition to such interactions mentioned earlier, there is possibility of interaction of an π- electron of one graphene layer to an π- electron of another graphene layer. The distance between the two layers of graphene is 0.3354 nm (Figure 4.22Y), which is larger than the distance of 0.142 nm (Fig 4.22Y). Therefore, the band formation due to such overlapping would be smaller than the band formation due to C-C carbon present within the individual layer. This picture becomes very different if multilayer graphene contains the structure as shown in Figure 4.22Z . If graphene layers have a

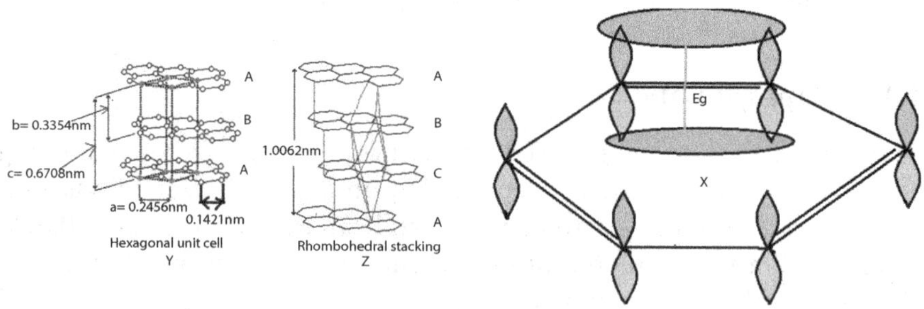

Figure 4.22 (X) Sketch showing the overlapping of sp^2 electron (i.e., overlapping of electron of π–π levels) (Y) Distance between C-C of different position in a hexagonal unit cell and (Z) distance between C-C of rhombohedral unit cell.

rhombohedral structure, the C-C distance between the first and third layer would be 0.10062 nm, which is even larger than all of them. Such a graphene layer would show a band gap much smaller than other band gaps. Considering such complexity in the structure of graphene and number of layers present in a given graphene sheet, there can be more than one type of band gap in addition to having a direct and indirect band gap. Experimentalists are needed to look into such complexities of graphene and determine the real experimental band gap of this material so that such mysteries could be understood.

4.5 Measurement of Mechanical Properties

Graphene-based sheets, which have outstanding properties due to their layer-by-layer hierarchical structures similar to the biomaterials such as bone and teeth, along with intra-layer strong sp^2 bonds and interlayer cross-links for efficient load transfer, instigate particular interest in measuring their mechanical properties.

Mechanical properties of this elite graphene material can be found by various measuring techniques such as Young's modulus, gas leak rates in graphene membranes, tunnelling of He atoms across a graphene sheet, classical effusion through single atom lattice vacancies and by extrapolating deflections and pressure difference. Some of the interesting techniques commonly used are discussed here.

4.5.1 Young's Modulus

Young's modulus (Tensile modulus or elastic modulus) measures the stiffness of material. This property is the ratio of the stress along an axis over the strain along that axis. But these values must fall in the range in which Hooke's law holds. The relationship between Young's modulus and the stress is expressed by Hooke's law for materials in one dimension, which states:

$$\sigma = E\varepsilon \tag{4.1}$$

where the stress σ is the force per unit area, E is Young's modulus, and ε is strain. Young's modulus has as its units of pressure the Pascal (Pa or N/m^2 or $m^{-1}\cdot kg\cdot s^{-2}$). It is important to note the direction in which it is measured because Young's modulus may differ depending on the direction in which the force is applied with respect to the material's structure. Raman spectroscopy has been used to measure Young's modulus of graphene sheet. When strain is applied to the graphene sheet, there is a shift in the Raman's G-band and 2D bands. After comparison of the measured strain with numerical simulation, Young's modulus has been calculated and reported to be 2.4±0.4 TPa and 2.0±0.5 Tpa for single- and bilayer graphene respectively. For this experiment, graphene sheet grown over a substrate is placed into a vacuum chamber to apply a pressure difference across the graphene membrane. In the vacuum chamber, a pressure difference across the graphene membrane is applied by pumping out the vacuum chamber, which makes graphene to bulge upward like a balloon (Figure 4.23A). On the top of the graphene sheet, a laser beam is focused to measure Raman spectra. Spectra is compared when measured under vacuum and when under pressure. One can get a shift in G-band (around 1580 cm^{-1}) and other 2D band (at around 2700cm^{-1}) from which it is possible (Figure 4.23B and C) to calculate the Young's modulus. It can be observed that when pressure is applied, the peak shifts towards higher values as compare to when it is under vacuum. A detail of the calculation of getting Young's modulus from Raman spectra is discussed by Jae-Ung Lee *et al.*

Table 4.1 shows typical Young's modulus for various materials.

4.5.2 Poisson's Ratio

Poisson's ratio is also referred as Siméon Poisson. When a material like rubber is compressed, in a direction perpendicular to the direction of applied force, it tends to bulge out or expand in a direction of 90 degrees to the applied force. This phenomenon is called the Poisson effect and is expressed as ratio of υ (nu), the amount of expansion to the amount of compression (Figure 4.24). This is a dimensionless quantity and is <1.

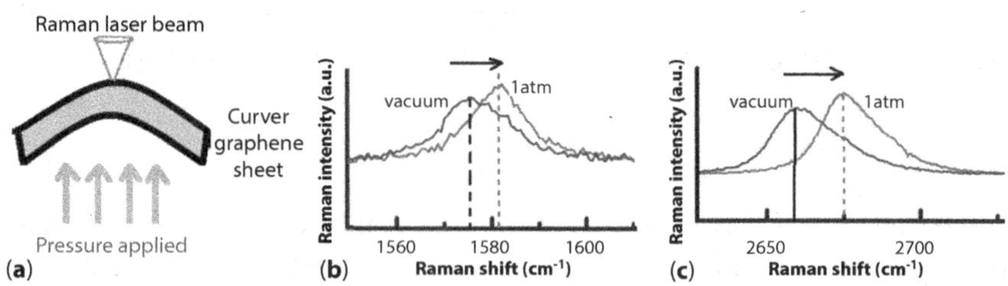

Figure 4.23 (a) Graphene sheet curved under a positive pressure (b) Raman shift obtained with graphene under vacuum (i.e., no pressure) and under pressure and (c) same Raman spectra but at around 2700cm^{-1}. Courtesy: Lee *et al.*, Department of Physics, Sogang University, Seoul 121–742, Korea; http://pubs.acs.org.

Table 4.1 Approximate Young's Modulus for various materials. Adapted from Wikipedia: Young's Modulus.

Material	**Young's modulus (*E*) in GPa**
Rubber (small strain)	0.01–0.1
PTFE (Teflon)	0.5
Nylon	3–7
Oak wood (along grain)	11
High-strength concrete (under compression)	30
Aluminium alloy	69
Titanium (Ti)	105–120
Copper (Cu)	110–130
Silicon (S)	150
Wrought iron and steel	190–210
Tungsten (W)	400–410
Silicon carbide (SiC)	450
Diamond (C)	1,050–1,200
Single walled carbon nanotube	1,000
Graphite/Graphene (within the plane)	1,000

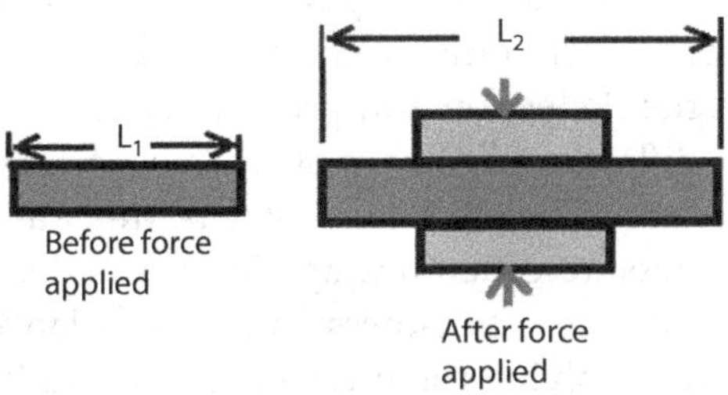

Figure 4.24 L_1 is the length of material before a force was applied to it and L_2 is the length after force is applied. Expansion occurs in direction perpendicular to direction of force applied. Poisson effect $\upsilon = -L_1/L_2$.

Table 4.2 Poisson ratio of some common materials. Adapted from Wikipedia: Poisson's ratio.

Material	Poisson's Ratio
Rubber	~ 0.50
Titanium	0.34
Copper	0.33
Stainless Steel	0.30–0.31
Concrete	0.20
Glass	0.18–0.3
Cork	0.00
Auxetics*	Negative

***Auxetics** are materials with a negative Poisson ratio, because when stretched, they become thicker perpendicular to the applied force. Auxetic materials can be single molecules or a particular structure of macroscopic matter.

If material is stretched for example in the x- direction (L_2) then its width will decrease in the y- direction (L_1). The ratio of L_1/L_2 would be the Poisson ratio. Its value for graphene has been reported -0.165. Typical Poisson ratios of some materials are shown in Table 4.2.

4.5.3 Bulge Test

A bulge test is normally done with thin film. Thin film is clamped and pressure is applied to deform the film (Figure 4.25). The center of deflection ∂ is measured by laser interferometry. Stress and strain caused on the film is calculated from the deflection and pressure applied to the film. In this method a pressure difference (∂P) is applied across a clamped circular film with a radius curvature of R and maximum deflexion z at the top of the film is measured. The pressure difference applies a well-defined and uniform force across the membrane of thickness t, which is balanced by the induced biaxial stress σ. After applying the pressure, some equilibrium conditions will be set up. Under this condition, force operating from underneath the film should be equal to force operating from the top.

Thus,

$$\partial p \, \pi R^2 = \sigma \, 2\pi \, R \, t$$

$$\text{Or } \sigma = (\partial P \, R)/(2t)$$

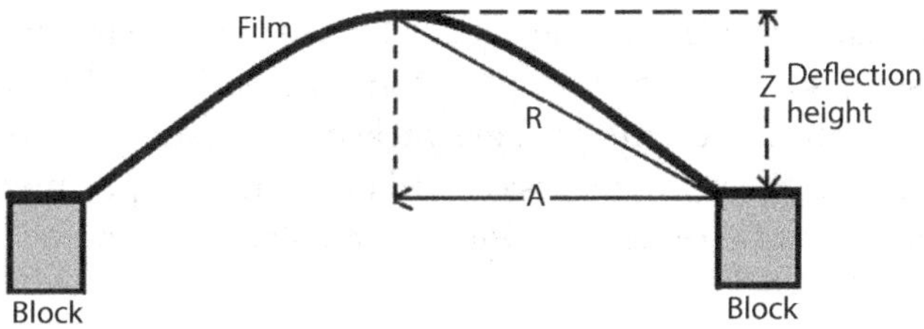

Figure 4.25 Schematic of bulge test.

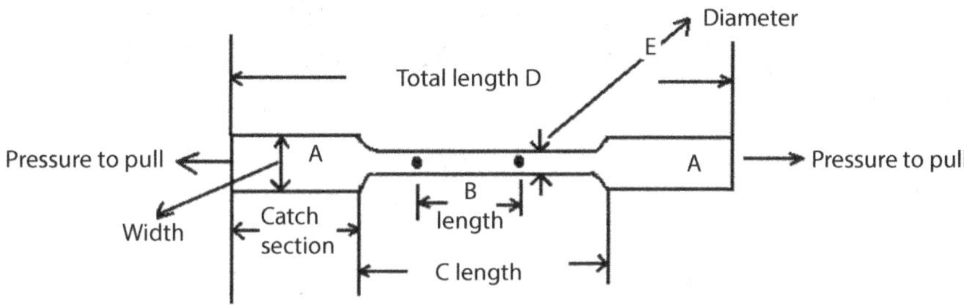

Figure 4.26 Description of test material and effect of its various changes with respect to applied pressure.

Bulge tests have been found to be important when graphene is being considered for use as a membrane. It is a reliable technique used to study the elastic property of graphene film.

4.5.4 Tensile Testing/Tension Testing

For Tensile testing (also known as tension testing), a sample is subjected to a controlled tension until failure. Among common tensile tests are ultimate tensile strength, maximum elongation and reduction in area. These measurements also help to obtain information about Young's modulus, Poisson's ratio, yield strength, and strain-hardening characteristics. It is another measurement to record tension of the bulged film.

By pulling sample (A) it is possible to quickly determine how the material reacts to forces being applied (Figure 4.26). When the material -A is pulled, it is possible to find about pressure the material can tolerate before it breaks. From the change in the length B, length C, catch total length D and diameter E, the entire behavior of the materials with respect to applied pressure can be calculated.

To test a material, the specimen A is attached with the machine through which tension is applied to the specimen. The variation in sizes A, B, C and D along with the diameter E are measured versus applied tension to specimen A. This measurement is carried out until the specimen fractures. The elongation measurement is used to calculate the *strain*, ε, using the equation 4.2:

$$\varepsilon = \frac{\Delta L}{L_0} = \frac{L - L_0}{L_0} \tag{4.2}$$

where ΔL is the change in length B, L_0 is the initial gauge length B, and L is the final length B when fracture occurred. The *stress*, σ, is calculated by using equation 4.3:

$$\sigma = \frac{F_n}{A} \tag{4.3}$$

Where F_n is the force and A is the cross-section of the length B. Finally, it is possible to draw a graph between stress and strain. Naebe *et al.* (2014) have studied the relationship between the stress and strain with graphene epoxy, functionalized graphene composite and epoxy (Figure 4.27). This graphs shows that this relationship is almost similar for all these three materials and it follows an exponential relationship. Functionalized graphene or graphene shows almost same behavior while epoxy shows slightly lower values.

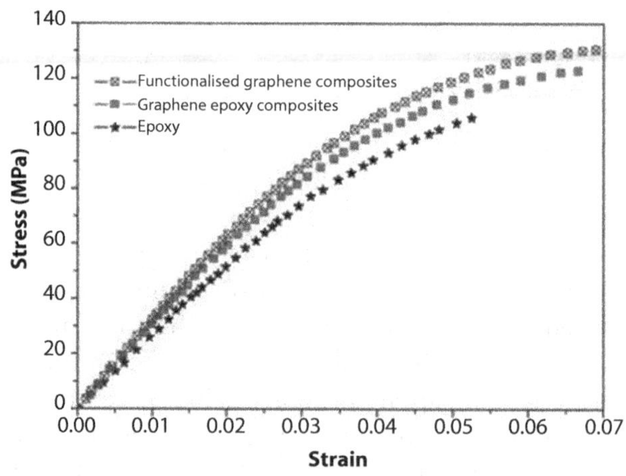

Figure 4.27 Graph showing the relationship between the stress and strain for epoxy resin, composite of graphene with epoxy and composite of functionalized graphene with epoxy. Copyright © 2013, Rights Managed by Nature Publishing Group, Courtesy: Naebe *et al.*, *Scientific Reports* 4, Article number:4375, March 2014, doi:10.1038/srep04375.

4.5.5 Gas Leak Rates in Graphene Membranes

Professor Sir Andre Geim reported that graphene oxide is impermeable to all gases including helium and liquids (vacuum-tight). But water diffuses through the membrane as quickly as it would do in the absence of any membrane. The diffusion of gases is measured by applying a pressure difference across the membrane. One-atom-thick graphene membrane makes it the world's thinnest unique separation barrier between two distinct regions. (Bunch *et al.* (2008).

4.6 Thermal Properties and Thermal Effect Analysis

This section includes the characterization of graphene by two processes.

4.6.1 Thermal Conductivity

As the name suggests, the property of a material to conduct heat is known as thermal conductivity and it is usually expressed in Kelvin-meters per watt ($K{\cdot}m{\cdot}W^{-1}$). Thermal conductivity of materials is temperature-dependent.

There are a number of ways to measure thermal conductivity. Each of these methods is suitable for a limited range of materials, depending on the thermal properties and the medium temperature. Thermal conductivity of suspended pristine graphene is measured by micro-Raman or Raman scattering spectroscopy. Using Raman scattering spectroscopy, Lee *et al.* (2011) monitored the temperature at the laser spot by the frequency of the Raman 2D band of the Raman scattering spectrum, and deduced the thermal conductivity by analyzing heat diffusion equations assuming that the substrate is a heat sink at ambient temperature. The obtained thermal conductivity values range from ~1800 W $m^{-1}K^{-1}$ near 325 K to ~710 $Wm^{-1}K^{-1}$ at 500 K. Thermal conductivity of some other materials is listed here. It can be observed from the table that thermal conductivity of graphene is much better than most of the materials. This is one of the reasons that graphene sheet is being considered for use as a heat sink in many electronic devices.

4.6.2 TGA and Thermal Stability

TGA (thermo-gravimetric analysis) can provide information about the thermal stability (Table 4.3), or the ability of the reducing agent to restore

Table 4.3 Thermal conductivity of some common materials.

Material	Thermal conductivity W/($m^{-1}K^{-1}$)
Silica Aerogel	0.004–0.04
Air	0.025
Wood	0.04–0.4
Hollow Fill Fiber Insulation Polartherm	0.042
Alcohols and oils	0.1–0.21
Mineral oil	0.138
Rubber	0.16
Polypropylene	0.25
LPG	0.23–0.26
Cement, Portland	0.29
Epoxy (silica-filled)	0.30
Epoxy (unfilled)	0.59
Water (liquid)	0.6
Thermal grease	0.7–3
Thermal epoxy	1–7
Glass	1.1
Soil	1.5
Concrete, stone	1.7
Ice	2
Sandstone	2.4
Stainless steel	12.11 ~ 45.0
Lead	35.3
Aluminum (alloys)	120–180
Aluminum 9pure)	237
Gold	318
Copper	401
Silver	429
Diamond	900–2320
Graphene	**(4,840 ± 440)–(5,300 ± 480)**

the crystal structure of the graphene sheet, which may be altered during the oxidation. We know that the graphite is a thermally stable compound. Its decomposition temperature is 600°C, whereas the decomposition temperature of graphene oxide is 200°C (Marconi *et al.* 2010), this is because oxygen atoms that have been added to the sheets of graphene oxide. If GO is reduced, i.e., oxygen is removed from the surface then the decomposition temperature increases. This increase depends on the reducing agent that is used. If decomposition temperature increases to 600°C, graphene should be similar to that of graphite. TGA of graphite, graphene oxide and reduced graphene is usually carried out under N_2 flow using thermogravimetric analyzer equipment. In Figure 4.28, TGA of pure graphite, graphene (reduced graphene) and graphene oxide is given. It is noticed that graphite shows almost a negligible change in mass for temperatures between room temperature and 600°C. Reduced graphene shows a linear decrease in the weight, whereas graphene oxide shows a rapid weight loss up to 200°C, and then it shows gradual decrease in weight loss up to 600°C. TGA is thus a useful technique to establish the purity as well as the stability of material.

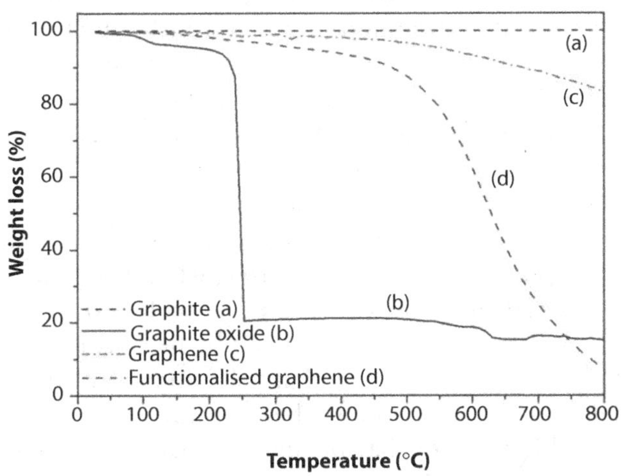

Figure 4.28 TGA plots for graphite, graphene oxide and reduced graphene. Copyright © 2013, Rights Managed by Nature Publishing Group, Courtesy: Naebe *et al.*, *Scientific Reports* 4, Article number:4375, March 2014, doi:10.1038/srep04375.

4.7 Characterization of Electrical Properties

4.7.1 Electronics

Graphene differs from most conventional three-dimensional materials. Intrinsic graphene is a semi-metal or zero-gap semiconductor. The electronic structure of graphene can be understood by studying the band structure of graphite. Wallace realised as early as 1947 that the energy-momentum relation shows linear relation, especially for low energies, which suggested zero effective mass for electrons and holes. These electrons and holes are considered to be fermions, and six corners of the Brillouin zone are called the Dirac points.

4.7.2 Electron Transport

Experimentally, it has been observed that graphene possesses high electron mobility (15,000 $cm^2V^{-1}s^{-1}$) at room temperature. It has also been observed that mobility of holes and electrons in graphene is nearly the same. These mobilities are almost same in the temperature range of 10K to 100K. This is possible only if defects do not exist in the graphene at these low temperatures. However, if scattering proceeds by phonon interaction, the mobility at room temperature is reported to be 200,000 $cm^2V^{-1}s^{-1}$ at a carrier density of $10^{12}cm^{-3}$. Under this condition, intrinsic resistivity at room temperature is observed to be $10^{-6}\Omega cm$, which is less than the resistivity of silver. Graphene shows its work function dependence on the type of substrate used for its synthesis. Similarly, mobility of graphene electrons deposited over SiO_2 is 40,000$cm^2V^{-1}s^{-1}$. Conventionally it is expected that once a material is doped with some cations/anions/neutral molecules, substituted molecules cannot be removed by any physical process like treatment of heat under vacuum. In graphene, some scientists (Schedin *et al.* 2007) observed that doped materials like gaseous species (acceptors or donors) or even potassium can be removed from graphene by heat treatment under vacuum. An interesting point to note is that these dopants did show an impact on the mobility of carriers, but after heat treatment the original mobility was observed. This type of behavior suggests that doping of graphene can be done without substitution of dopant for the carbon atoms. This also means that such doping is done at the interstitial sites already existing in graphene, and these dopants present at the interstitial site can affect the mobility of carrier. Furthermore, dopants present at the interstitial site can be removed by suitable heat treatment. It can also be concluded based on these observations that in order to maintain the neutrality of graphene, the dopant must take the interstitial positions such that it creates no charge imbalance in graphene. This conclusion requires a thorough investigation, especially by examining the charges and the sites in which the dopant is present in graphene layer.

4.7.3 Electrochemical Redox

Graphene oxide can be reversibly reduced and oxidized by applying appropriate potential stimulus. Controlled reduction and oxidation in two-terminal devices containing multilayer graphene oxide films result in switching between partially reduced graphene oxide and graphene, a process that modifies the electronic and optical properties. Oxidation and reduction are also shown to be related to resistive switching.

4.8 Work Function

Although the work function of graphene under a given metal electrode is critical information for the realization of high-performance graphene-based electronic devices, relatively little relevant research has been carried out to date. In this work, the work function values of graphene under various metals are accurately measured for the first time through a detailed analysis of the capacitance–voltage (*C–V*) characteristics of a metal–graphene–oxide–semiconductor (MGOS) capacitor structure. In contrast to the high work function of exposed graphene of 4.89–5.16 eV, the work function of graphene under a metal electrode varies depending on the metal species. Experimentally it has been observed that the metal contact induces charge inhomogeneity in graphene, which results in differences in the work function of graphene under metal and bare graphene. The work function of graphene deposited over Cr/Au metal shows ~4.3 eV, which is close to the work function of Cr/Au, whereas the work function of pure graphene is 4.89–5.16 eV. The work function of graphene deposited over Pd or Au is ~4.62 eV regardless of the work function of the metal. A study of the gate voltage dependence on the contact resistance shows that the latter case provides lower contact resistance. (*Song et al. 2010*)

4.9 Anomolous Quantum Hall Effect

In 1985 Klaus von Klitzing discovered the concept of Hall Effect and was awarded the Nobel Prize for this discovery. If a material of thickness t and surface area a is exposed to a magnetic field (B_H) in the direction

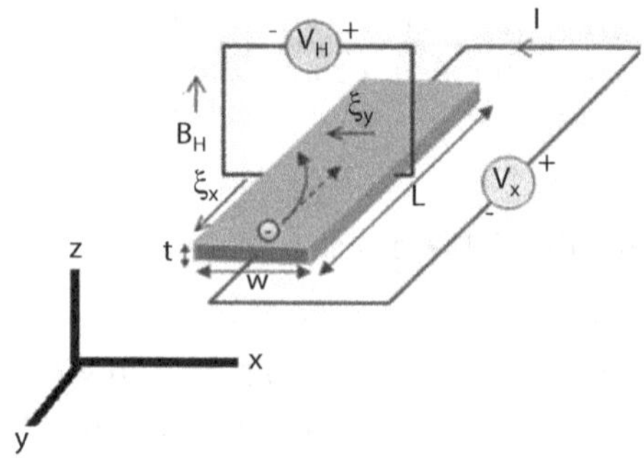

Figure 4.29 Schematic diagram showing the movement of electron carrier in presence of magnetic field. Courtesy: http://en.wikipedia.org/wiki/Hall_effect.

perpendicular to applied electric field (current I applied in y-direction), the carrier has been observed to move in the direction as shown in Figure 4.29, creating a potential (V_H) in the x-direction.

The Hall coefficient (RH) = ± 1/ne is observed with most of the semiconducting materials. The hall conductance is shown as σ_{xy} = ratio of Current (I) and Hall voltage and its quantized value is expressed as $\sigma_{xy} = \theta\,(e^2/h)$. Where θ is "filling factor," and is either an integer (ν = 1, 2, 3, …) or fractional (ν = 1/3, 2/5, 3/7, 2/3, 3/5, 1/5, 2/9, 3/13, 5/2, 12/5 ...) values. *e* is the elementary electric charge and *h* is Planck's constant. This effect is observed only with Si or GaAs solids but at a very low temperature around 3 K, and at very high magnetic fields.

Graphene, on the other hand, besides its high mobility and minimum resistivity, shows interesting Hall conductivity σ_{xy} in the presence of different magnetic field. Novoselov *et al.* have studied the Quantum hall Effect in graphene at room temperature, and have shown that the changes in the conductivity (σ_{xy}) occur for different values of ν and at different values of magnetic field. This remarkable behavior was measured at room temperature, at roughly 20°C.

4.10 Spin Transport

Graphene is thought to be an ideal material for spintronics due to small spin-orbit interaction and the near absence of nuclear magnetic moments in carbon (as well as a weak hyperfine interaction). Electrical spin-current injection and detection in graphene was recently demonstrated up to room temperature. Spin coherence length above 1 micrometer at room temperature was observed, and control of the spin current polarity with an electrical gate was observed at low temperature.

4.11 Summary

It is essential to characterize any material before the material can be used for any specific application. In fact it is the property of the material that suggests its possible applications as well allows for the understanding the behavior of material. In this chapter, emphasis was given on the characterization of graphene and its possible applications based on its particular character. The characterization was discussed by considering two basic broad approaches: microscopic and spectroscopic. Microscopic techniques are useful in understanding the morphology of the material. This is normally done by SEM, TEM, and AFM. Spectroscopic methods not only give information about the structural aspect of the material, but also its composition. These are accomplished by Raman UV-Vis spectrum, NMR, XRD, etc. In understanding the behavior of material, it is essential to know its optical properties, such as band gap, photoluminescence, etc. In certain applications of graphene it is essential to know its mechanical strength, which is measured by Young's modulus, bulge tests, etc. These aspects have also been discussed in this chapter. Two other important properties of material are its thermal and electrical properties. These two aspects of graphene have also been discussed. When graphene was discovered—for which a Nobel Prize was awarded—scientists thought that many scientific problems could be solved through the proper application of graphene. As a result many applications were proposed, but as to how many of those proposed applications will be economically viable, only time will tell. Nevertheless, in this section, some of the applications that the authors felt could be economically viable are discussed in detail. At some appropriate places, the authors included their own views about the graphene-like possibilities of direct and indirect band gap in graphene.

5

Engineering Properties of Graphene

Science is concerned with what is possible while engineering is concerned with choosing from among many possible ways, one that meets a number of poorly stated economic and practical objectives.

Richard Hamming

5.1 Introduction

The ever-increasing demand of tuning the properties of graphene to get optimal application in whatever field scientists desire has resulted in the accumulation of enormous data related to physical and chemical tuning of graphene. For example, Jiang (2013) could improve the performance of graphene as an active electrode material for supercapacitors and oxygen reduction reaction by creating in-plane holes and doping with nitrogen.

The capacity of carbon atoms (existing in vapor form or under intense conditions) to unite in a typical honeycomb-like configuration has formed the basic structure of graphene and all the other nano-forms of carbon, such as fullerene, carbon nano tubes, carbon nano fibers, carbon nano beads, carbon dots etc. It is a wonder that in nature, this form of carbon configuration (that is graphitic carbon) existed even before our efforts to synthesize them by different methods.

The properties that have been enhanced or altered by engineering graphene are magnetic, mechanical, electronic, band-gap, field emission and structural properties.

5.2 Engineering Magnetic Properties

Graphene is a two-dimensional network of hexagonal carbon atoms with sp^2 configuration. In this structure if one of the carbon atom is extracted or a dopant is added by replacing one carbon atom, then the site at which the carbon atom is removed will behave like a hole (i.e., positive charged site). Carbon atoms adjacent to this removed carbon will provide electrons, forming a cloud of electrons. Thus, due to the removal of one carbon atom, the structure of graphene will appear to possess a positive charge covered with a cloud of negative charge. This type of situation is developed in graphene due to the presence of π-bonds. These electron clouds behave like a microscopic magnet carrying one unit of magnetism, spin. It has been experimentally observed that this type of situation behaves like magnet and on application of potential it shows some remarkable behavior of making magnetic moments switched on and off. It is necessary to understand that such behaviors cannot be observed with diamonds structure because all its bonds are σ-bonds and has no π-bonds. This property of graphene has attracted scientists because of its weak spin-orbit interaction, the ability to control electronic properties by the electric field effect and the possibility to introduce paramagnetic centers such as vacancies and adatoms. Magnetism of adatoms in graphene is itinerant and can be controlled by doping, so that magnetic moments are switched on and off. Thus, carbon-based magnetic material would expand the technologies depending upon the magnetism. These materials have opened an avenue of research in the design of nanoscale magnetic and spin electronics devices like spintronics. In magnetic material, each micro-magnet allows information ('0' or '1'), which is stored in two magnetization directions ('north' and 'south'). This area of electronics is called spintronics. Spintronics has allowed scientists to develop transistor-like devices, where information is inscribed by switching graphene between its magnetic and non-magnetic states. These states are read either in the conventional manner by pushing electric current through or, even better, by using a spin flow. Such transistors have been like the holy grail of spintronics.

It is worth noting that graphene itself is non-magnetic, but magnetic properties can be observed by doping graphene with either certain impurities or by creating defects in the structure by irradiating graphene with electron or ions. By either of these processes, single-atom defects are created, which result in quasi-localized states at the Fermi level.

5.3 Engineering Graphene with Enhanced Mechanical Properties

Graphene sheet, when composed with different types of polymers, metal or metal oxides, has enhanced strength and property, which makes it suitable for some specific applications. Graphene sheet, though known to possess excellent mechanical strength, good conductivity and thermal stability, unless it is encapsulated in some other matrix, it is difficult to utilize these special properties of graphene. Scientists have found that making a composite with a suitable polymer can help to utilize these specific properties for special applications. The properties and performances of graphene composite polymer also depends upon the dispersity of graphene within the polymer matrix, bonding between the carbon atoms of graphene and functional groups of the polymer matrix.

There are various ways in which graphene sheet can be mixed with the polymer matrix, e.g., *in situ* polymerization, blending in a molten condition, etc. It would be difficult to discuss all possible ways by which polymer graphene composites have been made so far at the bench scale as well as at the commercial scale. For extra reading, a good review on this topic by Xiaoying Qi *et al.* (2013), which covers various types of composites made with graphene and their applications, should be consulted. Here, some specific examples are discussed. One of the common simple large-scale methods is solution mixing, provided that the solubility or dispersity of graphene in polymer is possible. For example, aqueous soluble poly (vinyl alcohol) can be mixed with graphene oxide (GO) and then polymerized. Organic soluble polymers like poly (methyl meth acrylate) (PMMA), polycaprolactanne (PCL), polyurethane, and polyaniline (PAN), have been successfully used to polymerize with rGO. In order to increase the solubility of graphene, efforts have been made to functionalize it so that it is soluble in some suitable solvent.

Efforts have been made to conjugate polymers with semiconductor and graphene sheets. For such polymerization, linkage has been established via alternating double, triple bonds, as well as with aromatic or hetero-aromatic rings. Such types of polymers have been synthesized to develop organic light-emitting diodes (OLED), photovoltaic cells, organic lasers, solar cells, super-capacitors, etc. graphene-PAN nanocomposite has been made by polymerizing aniline in suspension of GO or rGO in acidic solution. Such PAN nanocomposites materials have been used to develop electrodes for supercapacitor. rGO-PPy nanocomposites are biocompatible, have a porous structure, and good electronic and ionic conductivity. Because of these properties, such nanocomposites have been used as biosensors exhibiting good selectivity and sensitivity, especially for the detection of dopamine in human blood serum.

Mechanical strength enhancement of graphene for various applications has been achieved by various methods enumerated below.

5.3.1 Homogeneously Dispersing Graphene in Polymers

According to Stankovich *et al.* (2006), when graphene is dispersed into polystyrene (PS), the percolation threshold of only 0.1 vol% due to excellent dispersion of functionalized graphene in PS matrix is observed. Moreover, mechanical and thermal properties of parent matrix are also enhanced by addition of 1 wt% graphene. Apart from that, graphene transforms insulating polystyrene matrix into electrically conductive composite.

Furthering such efforts, Ramanathan *et al.* (2008) created polymer nanocomposites with functionalized graphene sheets, which provided good polymer–particle interactions. They tried two different polymers and showed a shift in glass transition temperature of over 40°C when poly (acrylonitrile), 1 wt% functionalized graphene sheet was added. However, when only 0.05 wt% functionalized graphene sheet was added to poly(methyl methacrylate), they found that glass transition temperature is only 30°C. A similar trend was observed when modulus, ultimate strength and thermal stability were recorded. They also confirmed that functionalized graphene sheet–poly(methyl methacrylate) showed better mechanical strength than those for single-walled carbon nanotube–poly(methyl methacrylate) composites.

Another trial for enhancing the mechanical strength of graphene is done by using either hydrophilic poly(vinyl alcohol) (PVA) or hydrophobic poly(methyl methacrylate) (PMMA) polymers. Here, simultaneously graphene oxide synthesis and adding polymer is done. By this method,

Putz *et al.* (2010) fabricated highly ordered, homogeneous polymer nanocomposites of layered graphene oxide by a vacuum-assisted self-assembly (VASA) technique. Here, the assembling components interact due to Coulombic attraction. These composites exhibit a high degree of order. Inter-sheet spacing in this method can be tuned by the use of polymer content in the interlayer gallery. The hydrogen bonding ability of the intercalating species impacts the structure modulus and strength. The addition of graphene to PVA improves stiffness by 1000% in comparison to pure polymer, well above the rule of mixtures (ROM), whereas stiffness of PMMA-based composites remained the same as ROM. However, tensile strength exhibited an increase of >1100% above the pristine polymer.

5.3.2 Chemical Cross-Linking

Another successful attempt in significantly enhancing the mechanical stiffness by 10–200% and fracture strength by 50% is achieved by chemical cross-linking graphene oxide paper with less than 1 wt % of Mg^{2+} and Ca^{2+}. These two divalent metal ions chemically interacted with the functional groups of the graphene oxide sheets. Moreover, the oxygen functional groups on the basal planes of the sheets and the carboxylate groups on the edges can both bond to Mg^{2+} and Ca^{2+} (Figure 5.1). This edge-linked m-carboxylate works against tensile force to enhance mechanical properties of the paper (Park *et al.* 2008).

Covalent cross-linking of graphene oxide nanosheets with borate has also exhibited an ultra-stiff graphene structure, with a maximum storage modulus of 127 GPa. It can be tuned by varying borate concentration (Park *et al.* 2008). Scientists achieved covalent cross-linking of graphene with borate by annealing that drives condensation reactions between borate and surface-bound hydroxyls, because hydrogen bonding is a weak link in cross-linking network. It was the covalent linkage that increased the mechanical stiffness up to 120 GPa. Moreover, they also demonstrated that graphene films can accommodate ~50 MPa of strain.

5.3.3 Hydrogenation

Xue and Xu (2010) have suggested that the atomic structures, binding energies, mechanical and electronic properties of graphene can be significantly modified by the hydrogenation and applied strain. They have offered a way not only to tune the binding of hydrogen on graphene in a controllable and reversible manner, but also to engineer the properties of graphene through

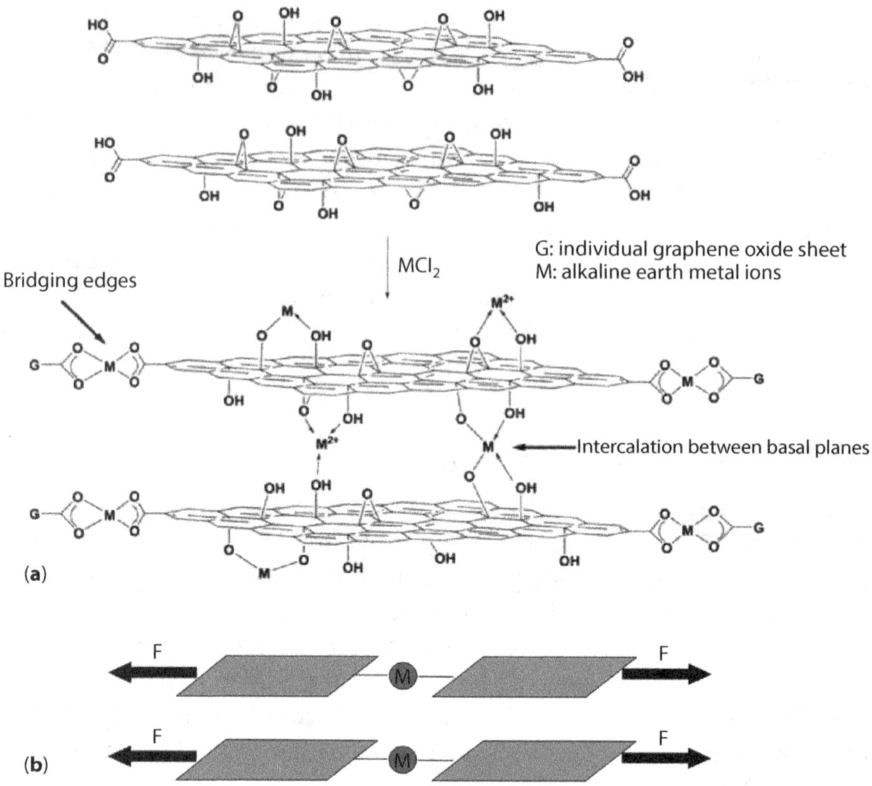

Figure 5.1 (a) Chemical cross-linking of graphene oxide paper with less than 1 wt % of Mg^{2+} and Ca^{2+} to increase the mechanical properties (b) Edge-linked m-carboxylate works against tensile force to enhance mechanical properties of the paper. Copyright © *ACS Nano* 2008; Courtesy: Sungjin *et al.*

a synergistic control through mechanical loads and hydrogen doping. Because monolayer graphene is only one atom thick, it is also amenable to external influences, including mechanical deformation. And mechanical deformation has remarkable effects on their binding strength, e.g., an in-plane strain of 10% can induce up to a 53.89% change in the binding strength in the symmetric phase and 23.56% in the anti-symmetric phase (also known as chair-like). In the anti-symmetric phase, the binding of hydrogen breaks the sp^2 characteristic of graphene, which can be partly recovered at tensile strain. Xue and Xu concluded that when hydrogen is bound to graphene, the energy gap opens up, thus reducing both stiffness and strength of graphene. This makes graphene a suitable material for hydrogen storage with tuneable and reversible controls from structural deformations and molecular doping synergistically. It has been observed that depending upon the extent of hydrogenation, either some or alternate carbon of hexagonal ring graphene changes from sp^2 configuration to sp^3 configuration. This type of change opens the band gap of graphene to as large as 2.54eV. This hydrogenated graphene is now classified as graphane.

5.4 Engineering the Field Emission (FE) Properties

Earlier in TV, a cathode ray tube (CRT) was used. This CRT, when very high voltage (to the tune of 2500V) was applied, it ejected the electrons, which falls on the fluorescent screen to produce an image. There were two basic disadvantages of this tube; it uses a very high DC voltage, and as a result, a large amount of heat is generated in the system. The main reason for applying very high DC voltage is the work-function of metal used for ejecting electrons. The work-function of most of the metal is in the range of 4.89–5.16 eV. Carbon material also has a work function in the range of 4–4.5eV, but due to its structural arrangement, its work-function can be reduced. For example, a cylindrical CNT possesses carbon at its circumference as well as on the sides of the tube. It has been observed that carbon present at circumference of CNT possesses almost zero work-function. This means the amount of energy needed to eject electrons from its circumference is almost zero as compared to from its side (Figure 5.2). Scientists have been trying to grow aligned CNT so as to make a large panel of vertically aligned CNT. Because it requires almost no DC potential, electrons can be ejected by applying potential as shown in Figure 5.2.

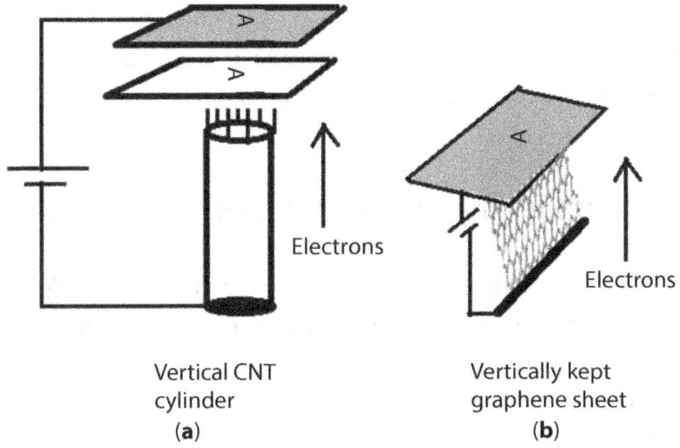

Figure 5.2 Schematic sketch to show the direction of flow of electrons when DC potential is applied to (a) Carbon nanotube and (b) vertically aligned graphene sheet.

After the discovery of graphene, scientists began research on whether vertically kept graphene sheet could be used for the same purpose, i.e., whether a CNT electron could be ejected from the side of a graphene sheet as shown in Figure 5.2B. Such a configuration was observed to show electron ejection from graphene sheet at very low DC potential. For commercial application of graphene for this purpose, it was thought to be desirable to make a graphene composite with polymer so that all graphene sheets were vertically aligned and the potential could be applied from one end of the graphene sheet. This was observed to be possible. This process of ejection of electrons from graphene or CNT is known as electron field emission.

Field emission (FE), field electron emission or electron field emission are the synonyms. The electrons are emitted from the surface of a material (as shown in Figure 5.2) into a vacuum, air, a fluid, or any non-conducting or weakly conducting dielectric, under an applied electric field.

The use of graphene sheet as field emitter faces a problem: It lays flat on the substrate that limits the field enhancement. To combat this problem, a composite of sheets of graphene with polystyrene, so that graphene sheet is oriented at almost a 90 degree angle so that it becomes somewhat vertically aligned with respect to the substrate surface (Figure 5.2B). This helped in field emission at low threshold fields (~ 4 V μm^{-1})

The graphene film exhibits different FE behaviours after adsorbing different gases. Hence, selective gas adsorption by graphene is used to engineer the field emission properties of graphene (Li and Sun 2014), for example, N_2 adsorption does not alter the FE properties of graphene, whereas O_2 or CO_2 adsorption increases the turn-on field, reduces the FE current and also affects the FE hysteresis properties of graphene films. Thus, graphene–gas adsorbate systems can be used for cold cathode material applications. It is an easy process where gas is adsorbed on the graphene by exposing it under reduced pressure.

5.5 Engineering Band Gap or Energy Gap of Graphene

A band gap of a material is the energy difference between the top of the valence band and the bottom of the conduction band. Band gap controls the electrical conductivity of a solid. Monolayer and bilayer graphene have zero band gap; therefore, it behaves like a metal. Because graphene is made of 100% sp^2 carbon, it shows no band gap. However, if some of sp^2 carbon is converted to sp^3 carbon, then it is possible to introduce a trigonal structure into the hexagonal structure of graphene. Such an introduction opens the band gap of graphene. It has been observed that if all alternate carbon atoms are converted to sp^3 carbon, then its band gap increases to around 2.5eV. There are various methods by which such conversion can be achieved, which are discussed below. In other words, graphene can be made to behave like a semiconductor if a band gap is introduced.

i. By applying an electric displacement to the two layers. By engineering a band gap in bilayer graphene, its energy gap can be finely controlled from zero to 250 meV. This electrical field has immensely increased the application of graphene as a flexible tool for nanoscale electronic devices.

ii. By chemically adsorbing one of the layers with adsorbed metal; this chemical bond (i.e., bond between the adsorbate to graphene sheet) opens a band gap in graphene. However, it must be mentioned here that there are chances of destroying the unique electronic structure of graphene, if it is chemically modified.

iii. By hydrogenation or chemisorption of hydrogen atoms. Most of the band-gap engineering of graphene has been done on GNR. Han *et al.* (2007) have demonstrated that the energy gap can be tuned during the fabrication process by controlling the width of the graphene nanoribbon (GNR), making it a suitable material for graphene-based electronic devices.

    Zheng *et al.* (2012) have engineered the band gap of GNR by di-hydrogenation of arm-chaired GNR (AGNR) edge, because di-hydrogenation reduces the effective GNR width. In mono-hydrogenation, a single H atom is used to saturate the dangling bonds and to make the system stable, whereas di-hydrogenation takes the carbon edges away and thus reduces the effective ribbon width.

iv. By classical doping with nitrogen or boron atoms

v. By physisorption of molecules with specific electronic structure, for example, physisorption of perylene-3, 4, 9, 10-tetracarboxylic-3,4,9,10-diimide (PTCDI). Physisorption is bonding by dispersive interactions and not by electronic interactions of the adsorbate orbitals with the graphene bands; hence it does not cause any significant modifications of the band structure of graphene. But if the lowest unoccupied molecular orbital (LUMO) of the adsorbate energetically lies at the energy of the Dirac point, a band gap is introduced in graphene.

vi. Adsorption of PTCDI induces a band gap opening on graphene, due to the specific electronic structure of PTCDI, characterized by a lowest unoccupied molecular orbital (LUMO) energetically located in the vicinity of the Dirac point.
This can be explained in terms of perturbation theory. According to perturbation theory, the effect of an interaction between the LUMO of the adsorbate and the graphene bands depends on the interaction strength, which is weak, divided by the energy difference of the LUMO and the graphene states.
For graphene bands that cross the position of the LUMO, this difference is approaching zero, and even a small interaction has an effect such that a crossing between the LUMO band of the adsorbate and the graphene bands is avoided.
In most cases these avoided crossings occur far from the Dirac point and have no significant effect. Only if the LUMO band crosses the graphene band in the immediate vicinity of the Dirac point, a band gap opens.
A similar effect can be assumed to occur if the band corresponding to the highest occupied molecular orbital (HOMO) crosses a graphene band close enough to the Dirac point.

5.6 Engineering the Electronic Properties of Graphene

Graphene is known to have remarkable electronic properties, basically due to a massless chiral character of charge carriers; and because of this, it exhibits an unusual metallic conductivity in the limit of no carriers and the half-integer Quantum Hall Effect (QHE) even at room temperature (Geim and Novoselov 2007; Novoselov *et al.* 2007; and Castro *et al.* 2009).

Because graphene is a zero-gap semiconductor, it has limitations in the applicability of graphene for electronics, where band gap plays a major role.

Nevertheless, as mentioned in chapter 2 it has been observed that graphene has the highest electron/holes mobility (500,000 $cm^2\ V^{-1}\ s^{-1}$), highest carrier concentration ($10^{13}/cm^2$), extremely good thermal conductivity (5000Wm K^{-1}), reasonably high sheet resistance (10–30 Ω/Υ)

and very high Young's modulus (1.1TPa). These are basic properties of graphene, which has made it a unique material for designing or engineering electronic properties.

5.6.1 Engineering Electronic Properties of Graphene for its Application in Transistors

In many electronic circuits, one component is connected with another component with the help of conducting metals, like copper. Although copper is a good conductor, it is not so good that is does not create some resistance. As a result, the printed circuit gets hot. Therefore, scientists like to replace the metal with graphene sheet so that it creates almost no resistance; thus creating negligible heat. In addition, most electronic circuits are fabricated over some polymers, which are not good thermal conductors. It is proposed to modify the polymers with graphene so that heat is dissipated faster, preventing the electronic circuits from accumulating heat and spoiling the life of electronic circuit. It has also been proposed to use graphene where metal has been used. For example, in designing a Metal Schottky junction, in place of metal it is possible to use graphene because the mobility of holes and electrons are almost same. Graphene has a very large surface area (2630 m^2 g^{-1}); wherever there is a need to increase the surface area, this material is taking preference. For example, in making a supercapacitor, graphene is much preferred because it provides a very large surface area and very high electron/hole mobility. Graphene can also be functionalized, which alters its properties of conduction as well as chemical reactivity. Therefore, it has also found its place in developing biosensors to detect various chemicals like DNA, glucose in blood, etc. Whenever requirements are for a high mobility of carrier, high thermal conductivity and large surface area in making any engineering gadget, graphene finds its place. This section contains specific examples of where graphene has found its place in electronic gadgets.

In graphene multi-terminal devices, graphene nanoribbons (GNR) are preferred as the wires can connect different nanodevices. Graphene's electronic properties are modified for many important applications, such as solar cell, supercapacitor, patterning graphene, making field-permeable graphene electrodes, intrinsic transport properties, piezoelectric properties, etc. Each of these applications needs a different type of modification of electronic properties.

In the following paragraphs, we will discuss how electronic properties are engineered. Much work in this direction has been focused on theoretical model to derive the possibility of engineering the electronic properties.

However, in this chapter we will be discussing the experimental efforts. Taking a clue from the many reports that breaking of sub-lattice symmetry in graphene produces an energy gap at the former Dirac point, Lee *et al.* (2013) have recently synthesized graphene sheets decorated with ultra-thin, Si-rich two-dimensional (2D) islands (i.e., Gr:Si sheets). They could modulate the electronic property of graphene by coupling (i.e., bonding to graphene via van der Waals interactions) with the ~2 to 4 nm thick and a lateral size of several nm Si-islands. They demonstrated that as compared to bare graphene, the Gr:Si sheet based field-effect transistors (FETs) exhibited enhanced trans-conductance and maximum-to-minimum current level. Moreover, their magnitudes gradually increased with increasing coverage of Si layers on the graphene. They further measured the temperature dependent current-voltage of the Gr:Si sheet, which showed a twofold increase in the resistance by decreasing the temperature from 250 to 10 K, thus confirming the opening of the substantial bandgap (~2.5–3.2 meV) in graphene by coupling with Si islands.

Before graphene could be used for some electronic devices, it is necessary to develop a way to reduce graphene's electron conduction. Twisting in graphene layer has given some hope to develop this property. On a graphene sheet, if properly formed, all layers of graphene will be parallel to each other such that the distance between each layer is maintained at 0.342nm. This would also mean that the distance between π-orbital (for example P_x or P_z orbital) of carbon of the top and bottom layer would also be the same. This would also mean that the electrical properties of graphene sheet, like electrical conductivity in-plane perpendicular to "ab" plane or parallel to "ab"plane, its band gap, and so on, would behave similarly throughout the surface of the graphene sheet (Figure 5.3A). However, once the sheet is twisted, this distance would not be the same throughout. As a result, the distance between the π-orbital will also be different (Figure 5.3B). Such a variation will undoubtedly hinder electron flow. This will also cause the variation in the interaction of π-orbital of carbon of the top layer to the bottom layer's carbon. These variations will undoubtedly induce the deviation in the electrical properties from place to place on the surface of the graphene sheet.

This twisted arrangement will induce strain, causing a magnetic field in the graphene sheet, with widths of even hundreds of nanometers. Such twisting will strongly affect the band structures of graphene. Twisting thus can affect electrons' localization either at the edges or at the center, and would hence impact the nature and magnitude of current.

Therefore, efforts are being made to control the twist in graphene layer at a specific place so that some useful nano size dimension electrical circuits could be designed.

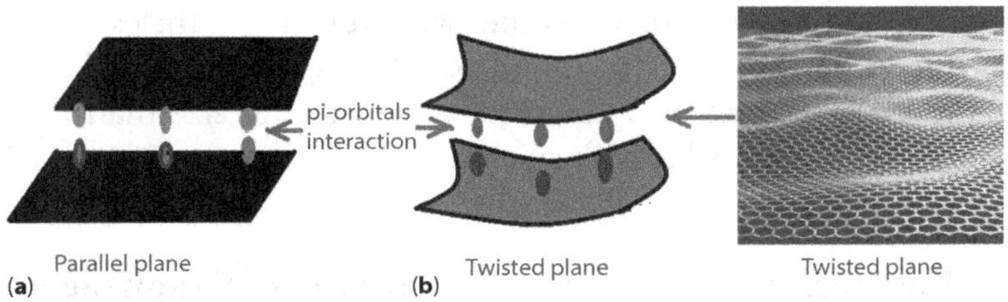

Figue 5.3 Schematic figure to explain the effect of twisting of graphene sheet on the π-π interaction.

Silicon or GaAs is used to make a transistor. These materials could not be made smaller than about 20 nm to make a suitable transistor in any electronic circuits. Scientists are looking for material that could behave like a semiconductor at sizes much smaller than 20 nm. The discovery of graphene suggested such a possibility. The major defect with graphene is that it is a zero band-gap material and unless a band gap is introduced into graphene, it is not possible to utilize this material where transport of electron/holes is required to be controlled by the magnitude of application of potential (i.e., magnitude of biasing potential). Recent reports suggest that band gap in graphene can be opened by doping with chlorine or controlled hydrogenation or by application of potential across a bilayer graphene.

The substitution of chlorine opens the band gap of graphene to a magnitude of 1.4eV while hydrogen of graphene produced graphane with band gap of 2.5eV. The opening of a band gap is possible because some of the sp^2 carbon of graphene is converted into sp^3 carbon. Recently it has been reported that an increase of applied current to a piece of graphene causes a drop in the voltage across the circuit. This drop in voltage is called negative differential resistance (NDR). This behavior can be used to perform logic, inverted transistor, etc. It suggests that a reverse phenomenon can also be possible, i.e., an application of potential across the bilayer graphene current should be produced in the plane of graphene sheet. The behavior of NDR is possible only if graphene shows the presence of band gap. NDR behavior cannot convert sp^2 carbon to sp^3 carbon as is observed by the doping process. Conceptually the formation of a band gap is considered due to interaction of electrons in the system. Since NDR is observed with bilayer graphene, this would suggest that NDR behavior must be altering the distance between the two layers of graphene, which must be altering the distance of separation of π- electrons of one layer to other layers. Therefore, there is a need to critically examine the effect of distance of separation between the two layers of graphene vis-à-vis current produced by the application of potential across the two layers. Nevertheless, whatever may be the cause of the creation of NDR, it has certainly opened up a new avenue to develop electronic circuit of much smaller size than even few nm, which was not possible with silicon.

5.6.2 Engineering the Electronic Properties of Graphene for Solar Cell Application

For solar cell application, graphene can be used either as a metal or as a semiconductor, according to how it is engineered.

5.6.2.1 p:n Junction Solar Cell

If graphene is to be used as a semiconductor in a solar cell, then one needs a semiconductor of n- and p-type with a band gap of 1.4eV (if solar energy is to be used for the illumination). Because graphene is a zero band gap material, it cannot be used to make p:n junction type solar cell. However, due to its more than 90% transparency, it can be used as a protecting surface of any inorganic material used for making p:n junction. In addition, since it has good mobility of carrier, graphene can be used as the collector of carrier in the p:n junction. Moreover, graphene has good thermal properties, so graphene can also be used as a base material of the p:n junction to dissipate heat from the cell. For this purpose, graphene is doped by different material, e.g., with:

i. *Chlorine*—added on both side of the surfaces of graphene to get p-type graphene (Figure 5.4A) with a band gap of 1.2eV. As much as 45% chlorine was doped on graphene by plasma-based surface functionalization technique (Wu *et al.* 2011), which yielded a carrier mobility of 1500 cm^2/V. By this process, the planar hexagonal surface is modified to a mixture of planar and tetragonal structure (Figure 5.4A). It is suggested that π- electrons are converted to σ-bond, which means converting sp^2 carbon to sp^3 carbon. Chlorine doping has been done by plasma technique, which means chlorine atom is converted to Cl^-. This ion while doping will create a hole at the site where the chlorine atom is substituted. This will result in the creation of p-type carbon. Converting zero band-gap graphene to 1.2eV band-gap carbon is most desirable to make a photovoltaic solar cell. It would be interesting to study the effect of replacing of all four hydrogen atoms by chlorine atom (Figure 5.4B). Band gap may increase with the number of chlorine atoms introduced into graphene, but the maximum reported value is 1.2eV.

ii. *Hydrogen*—Like chlorine, hydrogen substitution also has been tried on both sides of graphene, which has increased its zero band gap to 2.54eV. Hydrogen substitution causes alternate sp^2 carbon to be converted to sp^3 carbon. This type of graphene has been classified as graphane. The details of hydrogenation are given in chapter 2.

5.6.2.2 Schottky Junction Solar Cell

Metal-Schottky junction type solar cells need one metal as a current collector and a good conductor material of high mobility of carrier and almost zero band gap. Graphene meets these requirements, and hence, a graphene/Si type Schottky junction has been developed.

A graphene-Si (G-nSi) Schottky junction is fabricated, having the characteristics of a Schottky junction cell. The ideality factor, series resistance and shunt resistances are found to be 1.5, 10.5 Ω and 45MΩ, respectively. Open-circuit voltage (V_{oc}) of 0.42–0.48 V, a short-circuit current density (J_{sc}) of 4–6.5 mA cm^{-2} and a fill factor (FF) of 45–56%, corresponds to an overall solar energy conversion efficiency (η) of 1.0–1.7% (Li *et al.* 2010). Here, graphene sheet not only serves as a transparent electrode for light illumination, but also an active layer for electron–hole separation and hole transport. A Schottky junction solar cell with modified graphene films and silicon pillar arrays gives a conversion efficiency of up to 7.7%.

Moreover, to suit the requirements of a solar cell, electronic properties of graphene can be engineered by controlling number of graphene layer, tuning graphene work function and adding an anti-reflection film. As per theoretical conversion efficiency upto 9.2% could be achieved by such optimized graphene/semiconductor heterojunction solar cells. These types of solar cells will not be economically viable because one of the component of the cells, e.g., silicon will not be cheap.

5.6.2.3 Dye Sensitized Solar Cell (DSSC) or Organic Photovoltaic Cell (OPV)

In conventional OPV, ITO has been widely used as the electron/hole, because it is highly transparent (90%) to visible light, has low sheet resistance of 10–30 Ω/⊠ and suitable work function (4.5–5.2 eV). But there are many limitations in using ITO films; this film cannot be used to deposit semiconducting material needing a high temperature, unless ITO film is

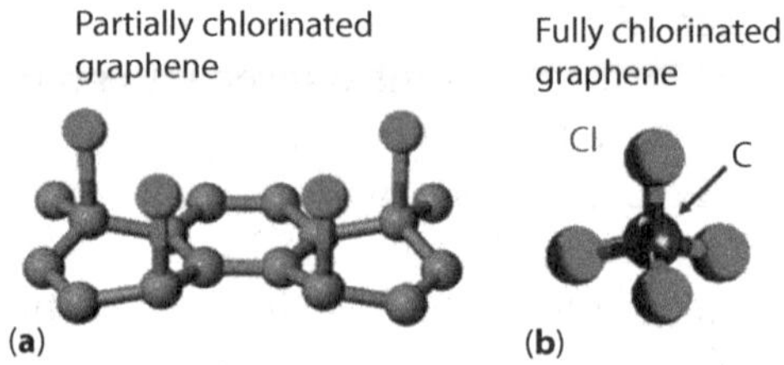

Figure 5.4 A schematic figure showing the formation of tetragonal structure in graphene due to substitution of chlorine atom (a) shows addition of four chlorine atoms replacing hydrogen atoms from four carbon atoms (b) shows how the tetragonal structure is formed by replacement of four chlorine atoms. Copyright © 2011, American Chemical Society, Courtesy: Wu *et al. J. Am. Chem. Soc.*, 2011, 133 (49), pp 19668–19671.

prepared on quartz glass, which makes it very expensive. ITO films are not stable in some specific pH levels, which limits its use in organic photovoltaic cells.

For DSSC, one needs a metal that is highly transparent, possesses a good mobility of carrier, zero band gap and suitable work function so that photogenerated carriers could be collected without creating Schottky type junction. Graphene has found its place for such development. As graphene has very similar properties as that of ITO, efforts are being made to use graphene instead of ITO film. For this purpose, the graphene film is deposited onto some metal like copper or Nickel, and then *in situ* semiconducting material is deposited over the graphene layer. Base metal acts like an ohmic contact for making a solar cell. However, several improvements are needed before graphene could be used as transparent and conducting electrode especially when making organic photovoltaic cells. Graphene, being hydrophobic, needs to be made hydrophilic so that it can be used with solution of organic dye or electrolytes. In addition, with the work function being 4.89–5.16 eV, it is necessary to manipulate its work function so that an active layer deposited over graphene creates ohmic-type contact. Moreover, the surface of the graphene layer needs to be functionalized without affecting its work function so that the active layer can be attached to the surface of the graphene layer.

5.6.3 Engineering the Electronic Properties of Graphene for Patterning Graphene

Patterning in making complicated printed circuit board is becoming very common these days. Scientists are trying to make printed circuit boards as small as possible so that they can write as many as 1000 circuits in one small chip of a few mm. Scientists are trying to explore whether graphene sheet could be made as a base material to write many printed circuits, because of its high conducting in-plane perpendicular to c-axis and insulating in direction of c- axis. Mobility of either hole or electrons is very high and at the same time it is thermally very good conductor. The problem is, how to create a printed circuit on the graphene sheet. One of the processes tried is selectively breaking C-C bonds. The energy needed to strip off an atom from graphene lattice is estimated to be 17eV, which means the energy required to cleave single carbon bond is 5–6 eV. For this purpose, various techniques, like irradiating with energetic electrons, chemical reactions, high temperature treatment, and nanoimprint lithography, have been considered. Lithographic techniques have been used on graphene nanoribbons of ~20 nm. In graphene it may be possible to create a line with as small as 0.5nm. To fabricate a graphene nano-mesh (Bai *et al.* 2010), a rectangular plate of copolymer thin film possessing cylindrical domains is formed over SiO_2 and then fluoride-based reactive ion etching and oxygen plasma etching process is done to remove polymer films and silicon oxide. Finally, the oxide mask is removed by dipping the sample into HF solution (Figure 5.5).

It will be difficult to discuss these ideas in much depth here, the following sources may be of interest to the reader: (Geim and. Novoselov 2007; Allen *et al.* 2010; Geim 2009, Geim and MacDonald, 2007; Katsnelson 2006; Park and. Ruoff 2009; Wei and Liu 2010 ; Feng ,*et al.* 2012).

5.6.4 Engineering the Electronic-Chemical Properties of Graphene for Supercapacitor

A supercapacitor that stores energy is comprised of porous electrodes, electrolyte, and a membrane separating them. The overall performance of the supercapacitor is determined by the physical properties of both the electrode and the electrolyte materials. The charge storage capability of these materials is usually evaluated by their capacitance, which is associated with the electrode/solution interface, which varies with the type of carbon and its conditions of preparation and usually has a value of 20 ~ 50 $\mu F/cm^2$.

Graphene is being considered for its application in supercapacitors. Taking a specific surface area of 2630 m^2/g for graphene as an example, its ideal attainable capacitance would be 200 ~ 500 F/g because the magnitude of capacity generated by supercapacitor depends on the surface area, pore size and the conductivity of the material. Surface area and pore size of graphene can be improved by chemical activation, like treatment with

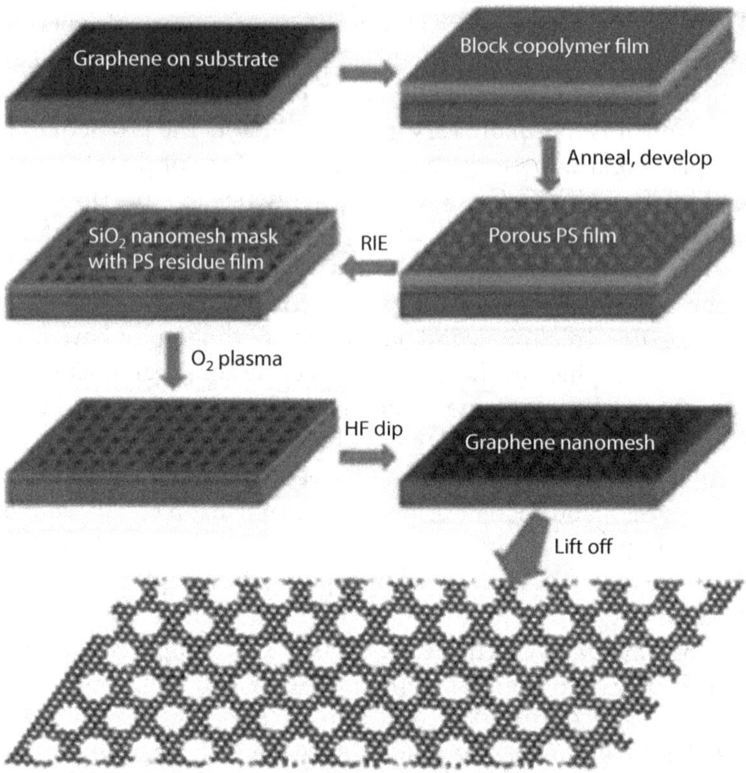

Figure 5.5 Schematic of fabricating graphene nanomeshes by block copolymer lithography. A block copolymer thin film with periodic cylindrical domains was first formed and annealed on top of a SiOx-protected graphene flake to develop a porous polymer film. Fluoride-based reactive ion etching and oxygen plasma etching were then used to punch holes into the graphene layer and remove the polymer film. Finally a Graphene nanomesh was obtained by dipping the sample into HF solution to remove the oxide mask. Copyright © 2010, Rights Managed by Nature Publishing Group; Reprinted with permission from ref. 48. Copyright[a] 2010 Macmillan Publishers Ltd: Courtesy: *Nature Nanotechnology*; 2010, 5, 190–194.

alkali. The conductivity can be improved by doping so that carrier concentration increases either by creating holes (positive charge) or electrons.

Recently, to improve the electrochemical performance of graphene, graphene with holes was synthesized by doping graphene with nitrogen, creating in-plane holes in its sheet plate and etching with KOH. These holes provides more active edge atoms, better accessibility to electrolyte, larger accommodation space for ions, faster electrolyte diffusion and movement (Jiang *et al.* 2014).

Another attempt to dope graphene with nitrogen (Wang *et al.* 2013) involves a hydrothermal method, using ammonium salts with hard acid-soft base pairs as nitrogen-doping agents. Thus prepared nitrogen-doped graphene exhibit enhanced electrochemical performance when used as electrode in supercapacitors, i.e., they exhibit a high specific capacitance (242 F g^{-1}) at a current density of 1 A g^{-1}, and remain a relatively high capacitance even at a high current density.

5.6.5 Engineering the Piezoelectric Properties of Graphene

When mechanical stress is applied to a piezoelectric material, it accumulates an electric charge. When the mechanical stress is removed, it loses its accumulated charge, regaining its original condition. Thus the piezoelectric effect is a linear electromechanical interaction between the electrical state of the material and the mechanical stress applied to it. Since this is a reversible system, the reverse can be performed with material to which if an electrical charge is applied, an internal mechanical stress is produced. Such materials are used for switching devices either to generate electrical signals or to produce internal mechanical stress, to produce ultrasonic sound waves, for production and detection of sound, generation of high DC voltage, micro balances, ultra-fine focusing of optical gadgets, and so on. It is also used in STM and AFM.

Graphene sheet has Young's modulus of 1.1GPa, which means it withstand high mechanical stress, and since electron mobility is very high, it will be possible to generate electrical voltage of high magnitude. The size of the sheet can also be small like graphene nanoribbons. Because the size of graphene nanoribbons can be made of a few nm, it is possible to fabricate nanosize piezoelectric graphene for various applications where other inorganic and some polymer-based piezoelectric materials cannot be used. The graphene sheet as shown in Figure 5.6 can be pulled apart and the DC voltage can be generated across the sheet. This phenomenon generates a new concept of "straintronics," i.e., electrical field strains like deformations in the lattice of carbon, producing a change in shape in anticipated habits. The concept of "straintronics" in graphene was established in 2010. It has been proposed that it may be possible to deliberately control the magnitude of deformation to graphene sheet as well as the site from where the deformation is desired to take place by the application of electric field, provided some specific cations/anions are inserted at the specific site of the carbon of the graphene sheet. If this is achieved it will bring promising implications for engineering.

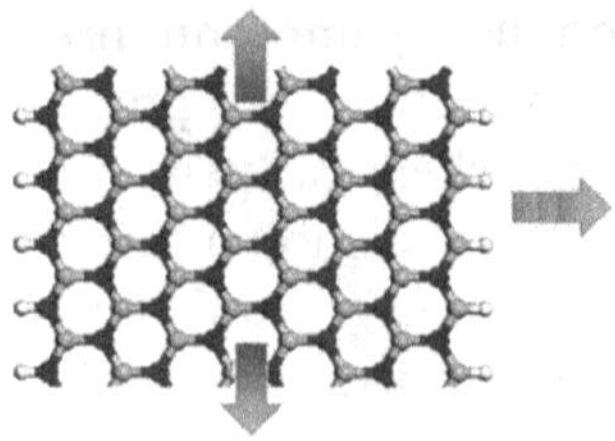

Figure 5.6 A schematic presentation to show how the applied mechanical stress in a graphene sheet can create electrical power in direction perpendicular to applied stress. Copyright © 2012, American Chemical Society; Courtesy: Oi *et al.*, *Nano Lett.*, 2012, 12 (3), 1224–1228.

5.6.6 Engineering Electronic Properties of Graphene for its Use in Fuel Cells

In the development of fuel cell operating with oxygen and hydrogen, one critical requirement is to develop an electrode for a reduction of oxygen without using expensive platinum. Platinum as an active electrode for oxygen reduction has long been used. However, platinum and other noble metal is not preferred because of its high cost, and electron transfer is not fast enough to give high power efficiency.

Nitrogen-doped graphene possesses good electrocatalytic activity for oxygen reduction. In addition, graphene can also be loaded with Fe or Co to enhance the electrocatalytic activity.

Nitrogen doping can be done by a simple method: Immobilizing Carbon nitride (CN on graphene sheets to form carbon nitride graphene (CN-G). For this purpose, $FeCl_3$ is added to CN-G and pyrolyzed. During this process simultaneously CN decomposes to tag nitrogen into graphene and Fe gets added to graphene. 5wt% Fe exhibits high electrocatalytic activity for oxygen reduction. Nitrogen doping can be done in the range of 4% to 12% by controlling the pyrolysis temperature (Figure 5.7) and nitrogen-doped graphene and its iron-based composite is an efficient electrocatalyst for oxygen reduction reaction (Parvez *et al.* 2012).

5.7 Engineering Structural Properties of Graphene

Graphene has potential for many applications, whether it is used alone or combined with other materials. Once the monolayer and bilayer or

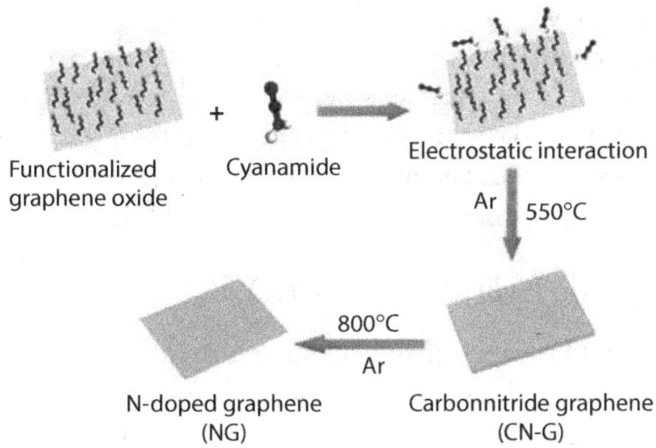

Figure 5.7 Preparation of nano-doped graphene (NG) for oxygen reduction reaction. Copyright © 2012, American Chemical Society; Parvez *et al.*, *ACS Nano*, Vol. 6(11) 9541–9550, 2012.

few-layer graphene syntheses was perfected, large efforts were soon directed towards altering the graphene structure for the benefit of its applications.

The efforts included engineering the:

1. Hybrid structures
2. Hetero structures (K)
3. Superstructures of graphene and
4. Introducing imperfections in graphene
5. Graphene–Metal Nanowire Hybrid Structures

5.7.1 Engineering Hybrid Structures of Graphene

Making a hybrid or composite of graphene has been thought of, because it offers significant advantages, depending on the material that has been used for hybridization of graphene. Moreover, the different hybridization material is selected to induce extra quality in graphene for a particular application.

5.7.1.1 Graphene Hybridized with SnO_2

As explained, the two-dimensional carbon sheet arranged in a honeycomb structure, i.e., graphene, has high surface area (theoretical value of

2630 m^2/g) (Peigney *et al.* 2001) for improved interfacial contact, excellent electronic conductivity and high mechanical properties; hence, the functionalized graphene sheet has been studied for direct Li-ion storage. This required improved capacity and cyclic stability of the anode materials, which is achieved by increasing the spacing between the graphene layers and forming a composite with SnO_2 (Yoo *et al.* 2008).

5.7.1.2 Graphene Hybridized with TiO_2

Self-assembled TiO_2/graphene hybrid structure for enhanced Li-ion insertion has also been developed (Wang *et al.* 2009). For this purpose, anionic sulphate surfactants were used. This assisted in self assembly of nanocrystalline TiO_2, rutile and anatase, with graphene. This also helped in the stabilization of graphene in aqueous solutions. The TiO_2/graphene hybrid has enhanced Li-ion insertion/extraction in TiO_2, with more specific capacity, doubled at high charge discharge rates, as compared with the pure TiO_2 phase, which was due to increased electrode conductivity in the presence of a percolated graphene network embedded into the metal oxide electrodes. This hybrid structure emphasizes the use of graphene as a conductive additive in self-assembled hybrid nanostructures to enhance high rate performance of electrochemically active materials and use of metal oxide TiO_2 as electrochemical active oxide material. The functionalized graphene sheets are a promising conductive additive for Li-ion battery electrode materials.

5.7.1.3 Graphene Hybridized with NiO

Hybrid structures are made by sandwiching Nickel oxide (NiO) nano particles between planar nano graphene sheets for electrochemical energy storage (Wei *et al.* 2011) by a multi-step assembly approach. To ensure uniform distribution of metal oxide NPs on the graphene sheet, graphene oxide is homogeneously dispersed in water and nitrate aqueous solution, followed by "mild" post-treatment. This hybrid structure results in a membrane-like form.

As shown in Figure 5.8, planar graphene nano sheet and metal oxide nano particles are tightly fixed and can avoid aggregation or stacking during electrochemical charge–discharge cycling, which effectively maintains the active surface and leaves stable and open channels for ion transport. Such hybrid structures are ideal for electrochemical energy storage and could be a solution for easy manipulation for various applications due to the membrane morphology because this structure acts as strain buffer to accommodate volume changes of the nano particles in a fixed direction,

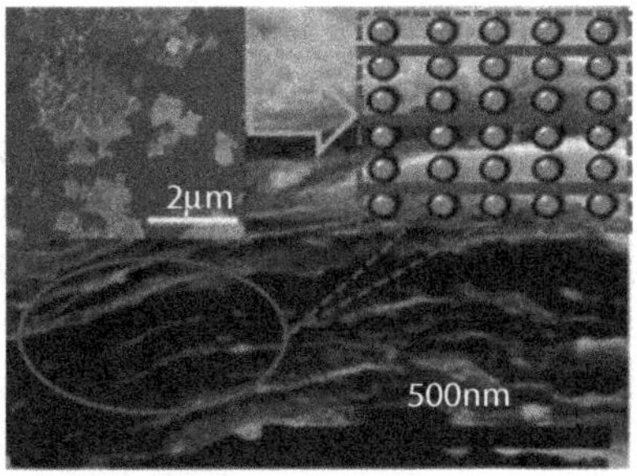

Figure 5.8 Hybrid structure made by sandwiching nickel oxide (NiO) nanoparticles between planar nanographene sheets for electrochemical energy storage. Copyright © 2011, Royal Society of Chemistry, Courtesy: Wei *et al. J. Mater. Chem.*, 2011,21,9014–9019.

and thus has a better resilience and structural stability in the electrochemical charge/discharge process.

Recently a sulphur, nitrogen-doped TiO_2/graphene oxide composite has been synthesized for high performance photocatalytic application, using thiourea (CS $(NH_2)_2$) as a binary-element doping reagent. The method involves colloidal blending. The photocatalytic degradation by S, N-doped TiO_2/GO was tested using methyl orange. As compared to photocatalytic degradation by TiO_2, this hybrid exhibits an average value of k (the apparent rate constant) to be ($k = 0.035\ min^{-1}$), which is eight times higher than that of TiO_2 grown in solution ($k = 0.0038\ min^{-1}$) and four times higher than that of P25 TiO_2 ($k = 0.008\ min^{-1}$).

The high photocatalytic activity of the composites is due to high specific surface area of graphene oxide, oxygen-containing functional groups and large aromatic domains that could bind by conjugated methyl orange molecules via π–π stacking. Moreover, S and N doping increases the numbers of photo-generated electrons and holes, which produce more charge carriers to form reactive species, and thus promote the degradation of methyl orange.

5.7.1.4 Graphene Coated with Transparent Thin Ferroelectric (P (VDF-TrFE)) Polymer

Though graphene has exceptional optical, mechanical, and electrical properties, its relatively high sheet resistance is a major constraint in its application for optoelectronics, photonics, and flexible transparent electrodes. To achieve low sheet resistance in monolayer form, graphene is doped up to $3 \times 10^{-13}\ cm^{-2}$ with a non-volatile ferroelectric polymer poly (vinylidene fluoride-co-trifluoroethylene) (P (VDF-TrFE) (Ni *et al.* 2011). This hybrid structure resulted in low sheet resistance of 120 Ω/ϒ at ambient conditions and more than 95% transmittance from the visible to the near-infrared range owing to the highly transparent nature of the ferroelectric polymer. Moreover, this hybrid of graphene has excellent mechanical flexibility and chemical inertness and can be used in transparent electrodes and optoelectronics.

5.7.1.5 *Graphene–Metal Nanowire Hybrid Structures*

This hybrid structure is fabricated by integration of two-dimensional graphene and network of one-dimensional silver nano wire. It has low sheet resistance (33 Ω/ϒ), high transmittance (94% in visible range), robust stability against electric breakdown and oxidation, negligible degradation of optical transmittance as well as superb flexibility (27% in bending strain) and stretchability (100% in tensile strain). The protection against thermal oxidation is due to the fact that silver nano wire is covered by a graphene layer. The hybrid electrode thus prepared presents superb mechanical flexibility and stretchability that enables it to completely fold in half. The multiple functionalities of the hybrid structures are being considered for applications in single-pixel displays integrated on flexible and wearable electronics (soft contact lenses) and hybrid electrodes to fabricate oxide semiconductor transistors.

5.7.2 Engineering Super Structures of Graphene

The super structure of graphene is mostly assessed by its moiré pattern (Figure 5.9 and 5.10), which is formed due to the lattice mismatch between graphene and substrate. Super structure is studied using atomic resolution scanning tunnelling microscopy (STM), X-ray photo-electron spectroscopy, Raman spectroscopy and the density functional theory calculation. Such studies have been done mostly by using Ru, Cu, Ir and Ni.

5.7.2.1 *Engineering Graphene Super Structure Using Ru*

Many attempts have been made (Marchini *et al.* 2007; Pan *et al.* 2007 and 2008; Vázquez de Parga *et al.* 2008, Sutter *et al.* 2008 and Jiang *et al.* 2009) to engineer Ru as substrate onto which graphene layer is deposited. For this purpose, commercially available Ru that contains some residual carbon also is used and annealed in ultra high vacuum. At an annealing temperature of 1000K to 1400 K, the carbon gets segregated from Ru and is deposited on the surface of Ru as a single layer.

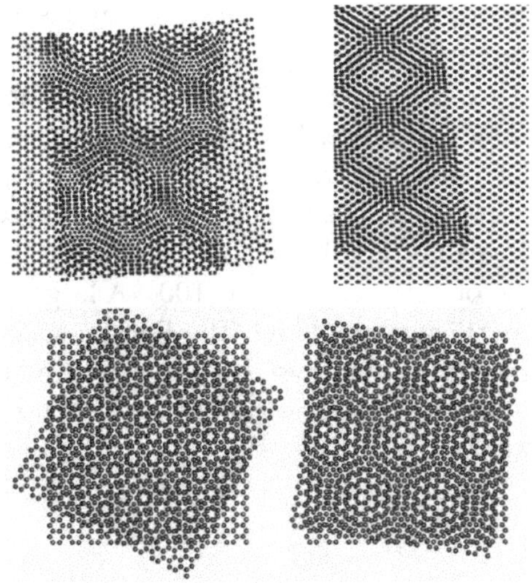

Figure 5.9 Various Moiré patterns of graphene. Copyright © 2011, Rights Managed by Nature Publishing Group; Courtesy: MacDonald and Bistritzer, *Nature*, 2011, 474 (7352).

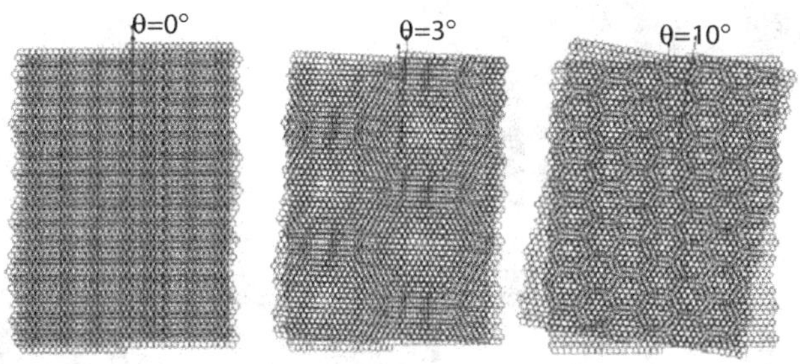

Figure 5.10 When graphene is twisted the Moiré patterns change like in a kaleidoscope showing different Moiré patterns

Recently, in an attempt to fabricate large-scale, high-quality and single crystalline graphene epitaxially grown on Ru (0001) substrate (Li *et al.* 2009), formation of moiré pattern of the graphene/Ru(0001), due to the lattice mismatch between graphene and Ru (0001) was noted. Using this super structure as a template, an organic molecule was assembled and the interaction between molecule and substrate resulted in engineering different superstructures. This may find many applications in the future.

5.7.2.2 Engineering Graphene Super Structure Using Cu

On single layer Cu crystal, when monolayer graphene is grown by passing a hydrocarbon (ethylene) gas in a CVD set up at 800–1100°C in ultra high vacuum (10^{-10} Torr) then as the STM shows the graphene grows primarily in registry with the underlying copper lattice for both Cu (111) and Cu (100). On Cu (111) graphene has a hexagonal structure, whereas on Cu (100) it has linear superstructure. The quality of the graphene film depends on the structure of copper substrate. The best growth occurs on hexagonal phase of Cu (111) surface because it matches with hexagonal phase of graphene. The domain size of graphene also becomes larger as compared to grown of other type of surface, e.g., Cu (100). A larger domain promotes the electronic properties of graphene. The Raman spectroscopic analysis is completed to understand the crystallographic orientation of the copper grains, which is needed to know the quality of graphene films.

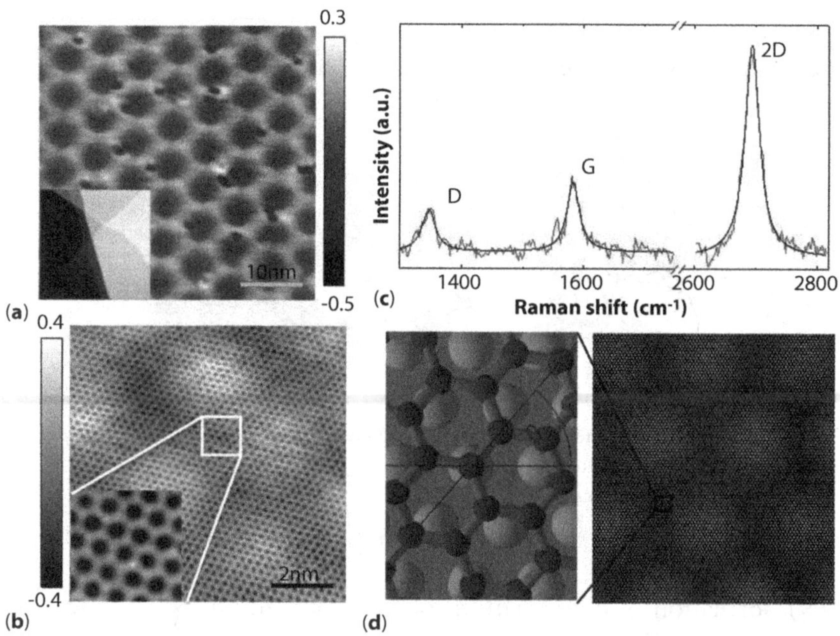

Figure 5.11 (a) Atomic terrace (V=0.6V, I=1.0 nA) after exposing clean Cu (111) to $3x10^{-5}$ L of ethylene at 900°C, showing a hexagonal superstructure across the surface, and point defects in the underlying Cu. (inset) 500x500 nm^2 topograph showing atomic terraces. (b) Higher resolution topograph showing the atomic scale structure of the graphene on the copper. (inset) honeycomb lattice of graphene. (c) Raman spectrum (after subtraction of Cu luminescence background) of CVD-grown graphene. (d) (left) schematic view of the overlay of graphene on Cu(111) (right) large-scale view of the resultant superstructure for the case θ=0. Copyright @ 2011Elsevier Ltd, All Rights Reserved; Courtesy: Zhao *et al.*; arXiv, 1008, 3542, 2010.

5.7.2.3 *Engineering Graphene Super Structure Using Ni*

Like Cu, polycrystalline Ni is also used for engineering graphene layers of controllable thicknesses and size by CVD process (Zhao *et al.* 2009 and Li *et al.* 2009). It is worth mentioning here that in this process, the morphology of graphene layer depends on the cooling rate after the pyrolysis has completed. In an attempt to understand the role of metal surface orientation on the morphology and structure of graphene, isolated three-dimensional islands of Ni film (~ 3 Å thick) with crystalline facets of different orientations (Murata *et al.* 2010) is used. Onto this Ni island, graphene is deposited. Highly oriented pyrolytic graphite (HOPG) is used as precursor and also substrate for Ni deposition.

Using the UHV-STM growth and structure of graphene on (111) and (110), facets of 3D Ni islands was investigated, which showed hexagonal and striped Moiré superstructures of graphene on (111) and (110) surfaces of Ni, respectively.

Based on the growth results, the possibility of single crystalline growth over large areas with potential technological implications for low-cost, large-scale fabrication of graphene on polycrystalline metal thin films or foils is clear. The metallic behavior of graphene on Ni surfaces is encouraging since the nature of the metal-graphene contact (Ohmic or Schottky) is important for graphene-based devices.

5.7.3 Engineering Hetero Structures of Graphene

The integration between graphene and another material is done in such a way that they do not form a hybrid, but rather produce a material where both materials exhibit their property and improvise the desired property of graphene. Because the monolayer of graphene is one atom in thickness and every carbon atom is a surface atom, it can interact with different atoms, molecules, nano particles, metals, semiconductors and insulators, thus forming graphene hetero structures. This is possible by replacing hydrogen atoms attached to monolayer carbon.

When the substituted metals are used as substrate, alterations in graphene structure due to a substrate-induced structural distortion, adsorbate, local charge disorder, atomic structure at the edges, and even atomic scale defects occurs, forming a hetero structure of graphene. These hetero structures of graphene are very important for the transport properties of graphene, specifically, lowered carrier mobility and suppression of weak localization in graphene-based devices caused by corrugation of the graphene.

5.7.3.1 *Engineering Graphene Hetero Structure Using Silicon dioxide (SiO_2) as Substrate*

SiO_2 is an insulator and is used as a support substrate for graphene sheet. This hetero structure is used in electronic devices. A combined SEM/AFM/STM study has revealed that in this device there is a break in hexagonal lattice due to the presence of a strong spatially dependent perturbation. The observation that graphene can conform to substrate morphology has opened new vistas to study how the corrugation-induced strain impacts the transport properties in graphene.

5.7.3.2 *Engineering Graphene Hetero Structures Based on Ultrathin Hexagonal Boron Nitride (h-BN)*

Graphene and h-BN have similar crystal structure and almost same lattice constants. However, graphene is a conductor, whereas h-BN is an insulator, having large energy band gap of 6 eV (Dean *et al.* 2010 and Gorbachev *et al.* 2011). Thick layers of boron nitride are used as substrate for many applications, for example, in flexible electronics, etc. This hetero structure is fabricated by sandwiching boron nitride between two layers of graphene (graphene/BN/graphene). The bottom graphene layer is fabricated by depositing narrow flakes of graphene on a SiO_2 substrate, and then atomically thin tunnel barriers are fabricated by micromechanically cleaving a few layer flakes from an h-BN crystal, which is then deposited on the graphene flakes using a dry-transfer technique. The top graphene electrode was transferred by the same method (Britnell 2012).

There are many other approaches to fabricate graphene and boron nitride hetero structures. To make a buried metal-gate field-effect transistor (FET), stacked hexagonal boron nitride (*h*-BN) and chemically vapor-deposited graphene hetero structure is used (Jain *et al.* 2013). Here a thin *h*-BN multilayer serves as both gate dielectric and supporting layer for the monolayer graphene channel.

Another planar graphene/h-BN hetero structure is fabricated by growing graphene in lithographically patterned h-BN atomic layers, which can be peeled off the growth substrate and transferred to other platforms including flexible substrates (Liu *et al.* 2014). To grow an *h*-BN layer, ammonia borane (NH_3–BH_3) is used as a precursor and is deposited on copper or nickel foils by the CVD method. The substrate is then annealed for 20 min at high temperature (800°C) and 800 m torr Ar/H_2 flow. The CVD furnace is heated to 1,000°C and simultaneously in another zone of CVD furnace the ammonia borane was sublimed at ~100°C and then carried into the reaction region by an Ar/H_2 gas flow kept at ~400 m torr. Growth time varies from 10–20 min. For patterning the *h*-BN, a mask of a 300 nm layer of chromium (prepared by photolithographic method) is deposited on top for protection, and then a pattern is made by a laser cutter.

5.7.4 Introducing Imperfections in Graphene

Graphene's strength is in its perfection. The strength of graphene is controlled by fracture toughness. Imperfection in graphene lessens its strength and it becomes brittle. Imperfection is due to Soliton. When multilayer graphene is fabricated, it produces some specific defects that affect its conductivity. These stacked layers of graphene, instead of being flat sheets of repeating carbon atoms, grow in ripples. These ripples are solitons, and through this path (or electrical highways) the electrons shoot from one end to the other end of the graphene sheet, giving rise to electronic one-dimensional channels. Much work has been done to describe the mathematics and theory behind the electrical properties of solitons and their occurrence in graphene. Efforts are underway to either get rid of solitons or to make a controlled one-dimensional electrical highway.

Though imperfections in materials usually degrade their performance, in the case of graphene, defects are being exploited to generate novel, innovative and useful materials and devices. Defects in graphene could be structural, topological, point defects, dislocations and grain boundaries, due to doping or functionalization and single-atom defect edges in graphene, e.g., vacancies/edge type. Defects play a key role in graphene's physicochemical properties.

Defects in graphene can be as small as a missing atom from the hexagonal lattice. The defect is measured as fracture toughness and crack growth. Moreover, the larger the sheet, the higher the probability of defects. These defects are bound to affect the handling, processing and manufacture of graphene. But the imperfections have certain advantages also.

5.7.4.1 Imperfections to Improve Graphene Sensors

Though graphene makes very good chemical sensors, it has been said (Salehi-Khojin 2011) that sensors are better if the graphene is worse. The type and geometry of defects of graphene determines the response of chemiresistors. Imperfections causing disruption to sp^2 configuration of graphene alters its mechanical, chemical and electrical properties. In graphene, a single-atom defect (e.g., vacancies and H-chemisorptions) induces a spin polarized defect state. Therefore, to enhance the chemiresistors' sensitivity, the line defects and edges are engineered. Salehi-Khojin *et al.* (2011) have noted that "Nearly-pristine graphene chemiresistors are less sensitive to analytes molecules, because adsorbate binds to point defects, which have low resistance pathways around them."

5.7.4.2 Engineering Single Carbon Atom Point Defects in Graphene to Induce Magnetism

Two types of defects have been found to induce magnetism in graphene, i.e., the hydrogen chemisorption defect (produced by the irradiation with high-energy particles) (Lehtinen *et al.* 2004) and the vacancy defect (induced by hydrogen plasma treatment) (Ruffieux *et al.* 2000). In both these defects, only one carbon atom is removed from the π conjugation network of the graphene sheet. The single-atom defects on the graphene lattice give rise to quasi-localized states at the Fermi level. The graphene lattice is the bipartite lattice. It can be viewed as two interpenetrating hexagonal sub-lattices of carbon atoms (labeled α and β).

Based on computational methods and models, it has been found that the itinerant magnetism is triggered by simple defects in graphene and are stable over the wide range of concentrations. The magnetic moments were $1\mu_B$ per hydrogen chemisorption defect and $1.12\ \mu_B$–$1.53\ \mu_B$ per vacancy defect depending on the defect concentration. The itinerant magnetism does not require the presence of highly reactive unsaturated dangling bonds. Both ferromagnetic and anti-ferromagnetic scenarios of the magnetic correlation are possible, with the second being more probable for truly disordered systems. The vacancy defects caused partial suppression of magnetic moments and the broadening of defect bands.

5.8 Summary

In this chapter, various properties of graphene have been touched upon. It is interesting to observe that magnetic properties can be inducted into the graphene by suitable doping. Mechanical properties of graphene can be altered by making suitable composites with polymer. The band gap of graphene has been adjusted by introducing a suitable dopant like chlorine or hydrogenation of graphene. With such modified graphene, it has been possible to develop different types of solar cells. Zero band-gap graphene has been used to develop dye-sensitized types of cell or organic flexible solar cells. Due to a variation in the Fermi level of graphene, it has been possible to utilize this property to study field emission. Some applications are discussed based on graphene's properties like high electrical conductance, its piezoelectric properties, thermal conductivity, high transparency to visible light, patterning, and so on. The dependence of the structure of graphene on the type of substrate used for its growth has also been exploited for specific applications.

6

Applications of Graphene

> *There is no such thing as a special category of science called applied-science; there is Science and its Applications, which are related to one another as the fruit is related to the tree that has borne it.*
>
> Louis Pasteur

Graphene's tremendous potential has spurred encouraging interest in scientists of all branches of science. This chapter discusses the inputs given in this new branch of carbon science for its various applications, ranging from the novel extension of conventional device physics to applied biological applications. The infinite possibility of applications of graphene is due to its novel properties like being ultra light but very strong and still very flexible, being fire resistant yet does not retains heat, a superb conductor but not even helium can pass through it! Because graphene is a comparatively new emerging field, initially most of the attention was diverted to reducing the cost of its large-scale fabrication and understanding its characteristics. These studies are still ongoing to understand graphene in depth. The CVD synthesis of large-scale graphene film by scientists of Hong's team (Chang and Hong 2009) in South Korea by CVD method on thin Ni substrate (wafer sizes up to 30 inches) triggered research on its applications. Scientists have now started looking at its applications. Very few applications have been commercialized. Many of them are at the verge of completion and some are also fantasies, which may become a reality in the near future.

6.1 Application Possibilities

6.1.1 High Specific Strength Related Applications of Graphene

There are reports of carbon nanofibres (CNF)–impregnated polymer composites, which are used in the production of lightweight and strong aircraft. Because graphene is much stronger and lighter than CNF and strong enough to withstand diamond cutters, it is being considered for use in wind turbine blades, flywheel, aerospace, aircraft, transportation and high power transmission lines, and so on. Graphene integrated into plastics such as epoxy is expected to not only replace the steel in the structure of aircrafts, thus reducing weight but also improving the fuel efficiency. It is even being envisaged that these characteristics may help in developing the body armor for military personnel and vehicles.

However, one application due to graphene's high strength has reached the commercial arena, as the YouTek™ SPEED tennis racquet series from HEAD is ready to make an appearance. This tennis racquet-shaft will be coated with an atom-thick layer of carbon atoms.

6.1.2 High Surface Area Related Applications of Graphene

The possibility of applications of graphene that has been envisaged due to its high surface area includes it use as adsorbent material, for catalysts, and so on.

A joint effort of Moscow State University and Rice Lab of Chemists has developed microscopic, atom-thick flakes of graphene oxide. These flakes are water soluble and can be easily produced in bulk. They have a remarkable ability to very quickly bind to both natural as well as manmade radioactive material from contaminated water and condense them into clumps of solids. Graphene oxide's large surface area, its capacity to adsorb compounded with very fast kinetics of sorption is the key for this application.

The trials and experiments were conducted by creating simulated nuclear waste containing uranium, plutonium, radioactive isotopes of the actinides and lanthanides and substances like sodium and calcium that could negatively affect their adsorption. The absorption capacity and duration for the removal of nuclear waste by graphene oxide was tested and compared with that of the commonly used bentonite clays and granulated activated carbon across a range of pH levels. The results prove graphene oxide to be better in the removal of radioactive contaminants. Another advantage in capturing radionuclide by graphene oxide is that it does not make it less radioactive, so it can be recovered by skimming off the solids and burning the graphene oxide. It has been demonstrated that graphene oxide burns very rapidly and leaves a cake of radioactive material that can be reused.

Thus in the very near future, graphene oxide can be used to clean up the contaminated site of Fukushima nuclear plants. Moreover, this technology will offer the chance to revive mining of rare earth metals.

6.1.3 Graphene for Electrical Energy Storage

Storing the energy in batteries and capacitors when it is not being used has always been a problem. The solution is to develop a supercapacitor or a battery for energy storage. A supercapacitor is an electrochemical device that in conjunction with batteries is used as an energy storage device. However, if high-capacity batteries deliver high power in a shorter period, the life of the battery reduces. For providing high power for a shorter duration, supercapacitors are employed, because supercapacitors can provide high power for short periods, such as power to initiate switching on any electric motor. A supercapacitor can store a high electrical charge and can be charged very quickly. Hence it is used in conjunction with a battery for many applications.

Though supercapacitors and EDLCs (Electrochemical Double Layer Capacitor) are 100 years old, they have been continually improved to enhance performance, in the wake of an increased energy storage demand, and efforts are ongoing.

Graphene, having elastic and mechanical properties, is being researched as an electrode material in a supercapacitor, as transparent conducting electrodes in a dye-sensitized cell, etc.

To replace the activated carbon that is currently being used in EDLC, single-layer graphene has already been used (Stoller *et al.* 2008), because single-layer graphene sheet has a much higher surface area than activated carbon. The monolayer graphene sheet was chemically modified for making an ultra-capacitor. This cell exhibited high electrical conductivity and

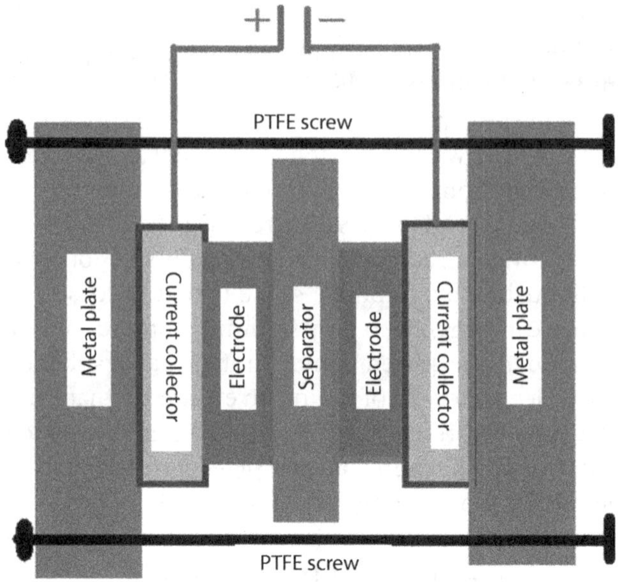

Figure 6.1 Schematic of a supercapacitor cell assembly.

specific capacitances of 135 and 99 F/g in aqueous and organic electrolytes, respectively. One typical sketch of a supercapacitor is shown in Figure 6.1 (Stoller *et al.* 2008).

It has been demonstrated that a supercapacitor can be developed for grid-scale integrated electrochemical charge storage, because graphene coating passivated surface charge traps and provides an ideal electrode-electrolyte electrochemical interface, thus improving the energy density by 40 times, and a two times wider electrochemical window as compared to unpassivated porous silicon electrodes.

It is expected that within the next 5–10 years, graphene-based micro-supercapacitors will be developed for use in both low energy usage applications such as lithium ion batteries, in tablet PCs (but at significantly lower size and weight), and graphene-enhanced lithium ion batteries for much higher energy usage applications, e.g., electrically powered vehicles.

6.1.4 Thermal Management by Graphene

The thermal conducting property of graphene has found applications at the nanoscale, microscale, macroscale for HVAC, giant computer servers, large area, transparent heating elements, etc.

The nanoflakes of graphene, due to their unique physical and thermal properties, were investigated (Rasheed *et al.* 2011) for their use in PCR (polymerase chain reaction). Graphene was found to enhance the PCR by increasing the yield of DNA by enhancing the thermal conductivity of base fluids. The PCR yield was found to be dependent on the size of nano graphene flakes.

6.1.5 High Flexibility Related Applications of Graphene

Graphene possesses high electron mobility as well as high mechanical strength. These two properties are suitable to fabricate flexible electronic circuits of various types. It is also possible to deposit graphene sheet over polymers as a base for making some electronic circuits like flexible screens, high-performance transistors and electronic components at low cost. If graphene sheet is to be deposited on a polymer, the preparation technique should not use temperature higher than 100°C. It would be even better at room temperature. For such purposes, techniques are developed to make graphene the solution of particles of size less than 100 nm. These can be stabilized in water by some suitable surfactants. This solution of graphene is used as conducting ink. This ink can be used as a single-layer sheet which can exhibit all the properties of graphene sheet and can be used to make graphene sheet on any type of substrates like glass, polymers like polyimide (a thermostable polymer), papers, etc. Such a unit is capable of working at very high frequencies in the GHz range.

In another method, graphene sheet solubilized in water is deposited on a substrate under an AC field applied between two electrodes. This method is also known as dielrophoresis, or DEP. AC fields guide the graphene deposition process so that at the desired specific place, a thick layer of graphene can be deposited. Such a thick layer of graphene helps in improving its performance at high frequency (8GHz). This type of transistor has many applications like a flexible screen, electronic devices built on different types of fabrics, etc.

There are several gadgets available for the detection of mercury in muscles of human beings in real time, but they are not very specific to Hg, and determination in real time takes a longer period. A liquid-ion gated field-effect transistor (FET)-type flexible graphene aptasensor with high sensitivity and selectivity for Hg has been developed. This graphene-based aptasensor has less than one second detection time. In this setup, graphene sheet is transferred onto a flexible substrate like polymer and this flexible sheet is used as aptasensor.

6.1.6 Electronic and Optoelectronic Devices Using Graphene

The use of graphene for logic and memory, transistors, integrated circuits for electronic applications, the attachment of functional groups to graphene aimed at self-assembly of simple circuits or the incorporation of chemical dopants to limit leakage current under zero gate bias is being researched all over the world.

Graphene is known to transmit more than 97% of light and offer electrical conductive properties exceeding 1 x 10^6 Ωϒm and therefore low electrical resistance. Therefore, optoelectronics is one area that is rapidly progressing, and soon there will be commercially produced optoelectronics, such as touchscreens, liquid crystal displays (LCD) and foldable organic light emitting diodes (OLEDs). These applications require better control of contact resistance. Though the indium tin oxide (ITO) is currently the most successfully and widely used material, graphene has been found to be potentially able to not only match the properties of ITO, but even surpass it. The optical absorption of graphene can be changed by adjusting the Fermi level, and high-quality graphene has a very high tensile strength and is flexible (with a bending radius of less than the required 5–10 mm for rollable e-paper).

Some of the success stories of the applications of graphene are in integrated circuits. Because graphene has a high carrier mobility, as well as low noise, it is used as the channel in a field-effect transistor. In 2008 the smallest one-atom-thick, 10-atoms-wide transistor was made; in 2009 an n-type transistor was made, and by 2010 IBM fabricated epitaxially grown graphene on SiC for mass production of integrated circuits. In 2011, IBM created the first graphene-based integrated circuit, a broadband radio mixer that can work at even at 127°C. And the year after that, in 2013 an eight-transistor 1.28 GHz ring oscillator circuit was made.

The electronic applications of graphene that are being envisaged include touchscreen and logic transistors. Because of high mobility, however, the issue of band gap–mobility trade-off must be looked into.

Graphene nanoribbons are being investigated for their application in FET transistors because graphene exhibits a pronounced response to perpendicular external electric fields. There have many patents regarding this since 2005. The basic FET made in 2004 has been improved by many since then by reversible chemical modification of the graphene layer.

6.1.7 Graphene as Lightweight Electrical Conductor

Graphene is a good electrical conductor in the direction of its growth plane (i.e., in the direction of ab plane) and insulator in direction perpendicular to its growth plane (i.e., in the direction of c-axis). This unique property can be used in spacecraft.

6.1.8 Transparent, Flexible, Conductive and Oxidation Resistant Films of Graphene

Graphene as a transparent, flexible and conductor material is expected to have huge implications for the PV solar-cell industry. It has found many applications for its use in image display and for coating metal as an anti-corrosive material etc.

Graphene in PV solar cell: Graphene has high electrical conductivity and optical transparency, absorbs a very low level of light, i.e., around only 2.7% of white light and at the same time has high electron mobility, and therefore can be used as an alternative to ITO. Moreover, it cannot be used in place of Si because graphene has almost zero band gap. Recently some reports have appeared stating that the band gap of graphene can be increased to 1.2eV by chlorine doping and even 2.54eV by hydrogenation. It has also reported that by doping with boron it can be made into a p-type semiconductor. Considering these possibilities, Wang *et al.* have developed a p:n junction n-CNT/P-Gr (Wang *et al.* 2014). This work will enhance the possibility of developing p:n junction with graphene without using Si either as a base or as a semiconductor.

Another advantage is that graphene films have high transparency in the visible and near-infrared regions, and they are chemically and thermally stable and have high charge mobilities, which recommends their use as a charge collector and transporter in photovoltaics (PV).

Due to its flexibility and ultra thin structure, graphene-based photovoltaic cells could be used to help mobile recharging or as retro-fitted photovoltaic window screens or curtains to save power.

Initial attempts were made in fabricating a graphene-silicon heterojunction solar cell. In this solar cell, graphene was used as a transparent electrode and n-type silicon was used to collect photo-generated carriers (Li *et al.* 2010). This cell has been improved by them by coating a wafer of silicon with chemically doped graphene with trifluoromethanesulfonylamide, which increased its efficiency to 9.6% (Li *et al.* 2010).

Some other successful attempts at research level are: nanographene-based photovoltaic cell that has shown > 12% efficiency, where semiconducting graphene is used as the photoactive material and metallic graphene is used as the conductive electrodes (Mukhopadhyay and Gupta 2013).

The other graphene photovoltaic solar cell with higher efficiency, i.e., 15.6%, is fabricated by using titanium oxide and graphene as a charge collector and perovskite as a sunlight absorber. Perovskite is a material with the same type of crystal structure as calcium titanium oxide ($CaTiO_3$). Though this device is made at high temperature (150°C) using solution-based deposition, it is supposed to bring down the cost by using flexible plastics (Wang *et al.* 2013).

To meet the commercial demand of graphene for PV cell, graphene is being produced by CVD method using methane as precursor and Ni as substrate. Onto this graphene, a protective layer of thermoplastic is laid. Finally, this plastic-protected graphene is attached to a flexible polymer sheet to be used in PV cells. The goal is to make a 150 cm^2 of this graphene plastic to make dense array of flexible PV cells.

There are certain challenges to be met in assembling graphene as a film electrode with low roughness, for example:

i. For graphene to be used in commercial solar cells, large-scale production is required. Peeling of pyrolytic graphene or thermal decomposition of silicon carbide are not very viable options for commercial synthesis of desired graphene, whereas graphene fabricated using solution processing contains lattice defects and grain boundaries, which act as recombination centers and decrease the material's electrical conductivity. Therefore, there is a need to fabricate a thicker or multilayer graphene sheet that can offer the desired resistance.
ii. The increased resistance can be combatted by adding silica as conductive fillers.
iii. Electrical conductivity of reduced graphene has been improved by attaching organic aromatic molecules like pyrene-1-sulfonic acid sodium salt and the disodium salt of 3, 4, 9, 10-perylenetetracarboxylic diimide bisbenzenesulfonic acid.
iv. Another hindrance is the zero band gap of graphene. To be used as a photo-active material, graphene should have a band gap 1.4–1.9 eV, if it is to be used for solar energy application.

Graphene for image display: Because of its more than 90% optical transparency and high conductivity, graphene is expected to be extremely good for transparent conductive electrode in flat panel displays and cathode ray tubes, touchscreens, etc. Moreover, graphene is a very strong material; hence, it is possible to make freestanding graphene to perform holography and coherent diffraction. Thus, graphene can be used for holographic imaging. This is because freestanding graphene can be transparent to low energy electrons as well (Figure 6.2).

Graphene as an oxidation-resistant coat on metals: Use of metals in many industries faces the corrosion problem due to reactive environments. One attempt (Chen *et al.* 2011) has shown that due to its oxidation-resistant ability, the graphene films grown by CVD method on the surface of the metal like Cu and Cu/Ni alloy protects them from air oxidation due to their chemical inertness to oxidizing gas and liquid solutions and by preventing the diffusion, even after heating at 200°C for up to 4 hours, though some partial oxidation could be noted on the grain boundaries. Moreover, graphene has been found to provide effective resistance against hydrogen peroxide. Therefore, it can be used on any metal that catalyzes graphene growth.

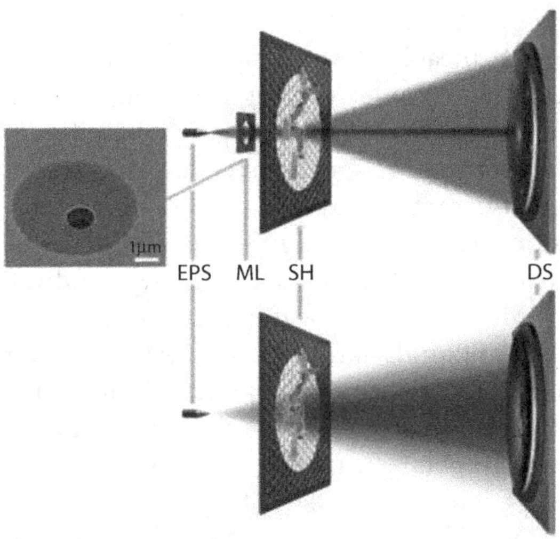

Figure 6.2 Experimental setup for coherent diffraction and holographic imaging of freestanding graphene. An electron point-source (EPS) emits a spherical electron wave. For coherent diffraction imaging (top), a micro-lens (ML) with a bore of 1 micron (see inset showing a SEM image of the lens) is employed to form a parallel beam impinging onto the freestanding graphene mounted onto a sample holder (SH). The diffraction patterns, respectively the holograms (bottom), are recorded at a 68 mm distant detector system, consisting of a 75 mm diameter micro-channel-plate, followed by a phosphorous coated fiberoptic plate and an 8000x6000 pixels CCD chip. Copyright © 2013, American Physical Society, Courtesy: Longchamp *et al.*, *Physical Rev Letters*, 110(25) 2013,255501.

There is a potential for graphene-based transparent conductive coatings to replace costly indium tin oxide-based transparent conductive coating, applied to energy saving windows, solar cells, display and LED, etc.

6.1.9 Graphene Film's Impermeability Related Applications

Only one-atom-thick, monolayer graphene membrane is impermeable to standard gases including helium (Bunch *et al.* 2008). Therefore, graphene can be used as an ultrafiltration medium. Hence, many applications of graphene have been envisioned based on this property, e.g., graphene-sealed micro-chambers to probe pressure in small volumes and explore pressure changes taking place during chemical reactions, phase transitions and photon detection. Moreover, in the future it may help in probing the permeability of gases through atomic vacancies in single layers of atoms.

Another application would be in the use of defect-patterned graphene membrane that can act as selective barriers for ultra-filtration and a unique separation barrier between two distinct phases of matter that is only one atom thick.

According to the observations of Geim's group (Nair *et al.* 2012), graphene oxide is permeable to water vapor, but impermeable to other liquid and gases. Using this property, Waugh has suggested distilling vodka at room temperature, without using any heating device for distillation. Thus, apart from the alcoholic beverage industry such membranes could reduce the production costing of biofuels also.

Graphene oxide sheets are biodegradable, and hence can be used as a permeable reactive barrier that will be useful in groundwater remediation.

One more standout property of graphene is that although it is completely impervious to liquids and gases, it allows water to pass through it. Therefore, graphene could be used as an ultra-filtration medium to act as a barrier between two substances used for molecular sieving.

An improvised graphene with nanometer-scale pores in single-layer free-standing graphene has the ability to effectively filter NaCl salt from water (Cohen-Tanugi and Grossman 2012).This suggests that graphene may be useful for the desalination of sea water.

Single-layer or multilayer graphene can form a protective film on a variety of materials as a barrier coating for gas and moisture. This is important to keep moisture/gas away from industrial gadgets to maintain their electrical characteristics for a long period. It is reported that 25 layers of graphene sheet can stop almost 90% of mercury vapor from transmitting. This experiment was used to conclude that food products could be coated with graphene sheets to prevent any toxic material from unfriendly environments to penetrate into the package and damaging the food.

The Bill and Melinda Gates Foundation awarded $100,000 to develop new elastic composite materials for condoms containing nanomaterials like graphene.

6.1.10 Reinforcements of Polymer Composites

The replacement of heavy weight steel by other material stronger than steel but much lighter than steel is needed. Though some polymers meet this requirement to some extent, their strength has not been very satisfactory. Graphene on the other hand possesses unique properties such as Quantum Hall Effect, carrier mobility (110,000 $cm^2V^{-1}s^{-1}$), large surface area 2630m^2g^{-1}, optical transparency (>97%), high Young's modulus (11TPa) and good conductivity (3000-5000 $Wm^{-1}K^{-1}$). Because of these superior properties of graphene, a composite of graphene with polymer can show good mechanical, thermal, electrical, flame retardant properties, etc. Therefore a composite of graphene with a suitable polymer can satisfy various requirements. However, these special properties of graphene composites with polymer largely depends on the distribution of graphene in the polymer matrix. It also depends upon the interfacial bonding between polymer and the graphene. Therefore, the properties of the composites depend upon the method of preparation. It is also reported that graphene oxide (GO) is well suited with polymers and it is preferred as nanofiller for making nanocomposites. Graphene is being applied for making electrically conducting composites, thermally conductive nanocomposites, for antistatic coating, ceramic matrix composites, transparent conductive composites, paper-like composite materials, etc. All these have unusual mechanical properties.

The glass touchscreen will be made using graphene as a conductor being replaced with plastic. Such touchscreens will be super thin, light, flexible and unbreakable. This will make an unbreakable cell phone as thin as a piece of paper and foldable enough to slip into a pocket.

6.1.11 Sensors

For single-molecule gas detection, sensors, lock-and-key type binding sites to achieve selective sensitivity to a wide variety of analytes, including chemical warfare agents or even biological species, is needed.

6.1.11.1 Graphene for Biosensors

Unique chemical, optical, electrical and electrochemical properties of graphene have prompted its use in and as a biosensor (Mohanty *et al.* 2008, Shao *et al.* 2010, Shan *et al.* 2010, Zhou *et al.* 2009) for enzymes (Shan *et al.*

2009 and Song *et al.* 2010), for DNA oligomers (Tang *et al.* 2010), DNA analysis (He *et al.* 2010 and Lu *et al.* 2009) for *in situ* molecular probing in living cells (Wang et al. 2010), detection of nucleic acid (Nelson *et al.* 2010), for thrombin detection (Chang *et al.* 2010). The efforts in this direction that have yielded fruitful results include strong binding of nucleotide bases in ssDNA, which was weakened after DNA hybridization to form dsDNA (Jang *et al.* 2010). Using this phenomenon and the fluorescence-quenching capacity of graphene, graphene-based DNA detection sensors are being developed (Balapanuru *et al.* 2010). Some examples include:

i. *For detecting DNA*—Because GO is water-soluble, GO-based multicolor DNA probes are being used for detecting DNA in solution. These probes are very rapid, highly sensitive and selective for detection of DNA in solutions (He *et al.* 2010).
ii. *For detecting bacteria, protein and DNA*—Devices for the detection of bacterium, DNA and proteins (Mohanty *et al.* 2008) have also been prepared.
iii. *To deliver oligonucleotides*—Graphene is being attempted for delivery of oligonucleotides (molecular beacons and aptamers) into living cells for *in situ* probing of biomolecules (Wang *et al.* 2010 and Lu *et al.* 2010).
iv. *For sensing of biomolecules*—Graphene-based composite materials are also being attempted in the eletrochemical sensing of biomolecules, e.g., glucose (Liu *et al.* 2010), DNA and proteins (Zeng *et al.* 2010). Owing to its ultra-high surface area and excellent electron mobility, graphene-based devices are very highly sensitive.

Although high detection sensitivities have been achieved by many graphene-based biosensors, there are many concerns that need to be looked into, such as batch-to-batch variations in those graphene biosensors, reliability and reproducibility.

However, graphene with large surface area, high electrical conductivity, thinness and strength, will surely help in the development of fast and efficient bioelectric sensory devices that can monitor bio-molecules like glucose, haemoglobin, cholesterol and even DNA sequencing, etc.

6.1.11.2 Graphene as Gas Sensors

Graphene has a large surface to volume ratio, large surface area and it can be doped with various cations/anions. These properties have made it possible to use graphene as a gas sensor. Graphene can adsorb gas in large quantity. The adsorbed gases change the concentration of surface states, which in turn change the surface resistance of graphene. Therefore, it is possible to measure the change in its resistance. It has been possible to extend the detection limits for example for NO_2 and NH_3 from ppm to ppb level. Sensitivity of graphene sheet has been improved by patterning the graphene so as to increase the surface active area of graphene and hence increase its detection limit.

6.1.11.3 Graphene for Chemi-Sensors

Graphene is applied as a chemical sensor due the following advantageous properties:

- Graphene or graphene oxide can detect chemicals up to ppm or ppb level due to its high surface-to-volume ratio.
- Electronic and mechanical properties can be tailored to perform the transduction of the sensing signal.
- Functionalization of graphene by polymers, metals or other modifiers helps to interact with specific molecules and in so doing improves the selectivity with acolyte.
- Mechanical robustness of graphene helps in not jeopardizing its two-dimensional delocalized transport properties.
- Graphene being highly conducting, a slight change in the interactions of molecules in tests causes a large shift in carrier (electron) concentration. As a result molecules at ppm or even at ppb level can be detected very accurately.
- Four probe devices can be used with graphene monocrystals. Thus the influence of the contact resistance can be minimized.
- Graphene and its oxide can be produced economically.

Intrinsic monolayer graphene has no dangling bonds on its surface. To improve the chemisorptions of target molecules on graphene surface, graphene needs to be functionalized with polymers, metals or other suitable modifiers to enhance the adsorption of target species, which can cause a change in electrical resistance to quantify the concentration of analytes.

It is interesting to observe that substrate onto which graphene is loaded plays a crucial role for tuning properties of graphene. For example, single-layer graphene loaded on SiC behaves like n-type with very low carrier concentration. This is because electron is donated from SiC to graphene.

Whereas multilayer graphene loaded on SiC behaves like p-type with a higher concentration of charge carriers. Another example is of Pd-loaded graphene. The Pd bedecked graphene was capable of detecting hydrogen from 0.5 to 1% in synthetic air while pure graphene is poorly sensitive to hydrogen. Therefore, it is necessary to examine the possible effect of substrate and its role in assisting the direction of flow of electron between graphene and the substrate. Moreover, it is also necessary to understand the role of adsorbing metal/ions or polymer on the graphene vis-à-vis its interaction with molecules in tests, i.e. whether the test molecule can form some specific compound with the material adsorbed on graphene.

6.1.11.4 Graphene for Pressure-Sensors

Graphene is a good material for nano-electro-mechanical systems (NEMS) due to its specific properties like high carrier mobility and high Young's modulus. It can be stretched to 20% without breaking. Graphene is also impermeable to gases including helium; hence it is possible to measure the gas pressure with the help of graphene sheet. By applying stress on graphene it can develop potential, which is a good property for its use as piezo-resistive material. This means that when pressure is applied to graphene sheet, it develops an electrical signal, which can be calibrated to determine the amount of pressure applied. For such purposes graphene is deposited over a suitable substrate like SiO_2, Silicon nitride and even on some polymers. It has been reported that sensitivity per unit area of graphene is 20–100 times more than conventional piezo-resistive sensor materials.

The sensitivity of piezo-resitive membrane-based pressure sensors is related to the applied pressure, resistance and applied voltage as per the equation -1.

$$S = [V_o - V_p]/V_o P \tag{6.1}$$

Where S is the sensitivity, V_o is the initial voltage applied, V_p is the voltage developed after applying pressure P. In this type of measurement, a constant current is applied to graphene sheet and due to its own resistance, some voltage is developed. This voltage is charged by applying pressure to the graphene sheet. Thus it is possible to develop a reference graph between the $(V_o - V_p)$ and the applied pressure.

6.1.11.5 Graphene for Strain-Gauge

Utilizing the piezo-resistive property of graphene, a strain-gauge sensor has been prepared. Graphene sheet is deposited over a suitable substrate

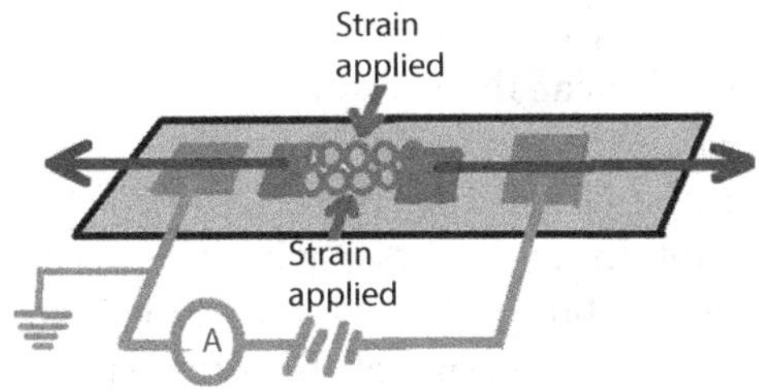

Figure 6.3 A schematic diagram of a strain sensor using graphene as a sensor.

like SiO_2. A schematic of graphene strain-gauge is shown in Figure 6.3. A strain is applied to a graphene sheet deposited over a substrate and the developed resistance is measured. A relationship between the change in the resistance (δR) and the mechanical strain (€) is measured. A common gauge factor (GF) is calculated which is defined as GF = (δR/R)/€, where € is mechanical strain. Gauge factor shows the sensitivity of electrical shift to mechanical deformation.

The detection of deformation or structural changes occurring in the infrastructure can be measured by such a device. This type of device can also be used to measure internal activities of human bodies. Graphene has become useful material for such devices because of its good mechanical property. Graphene-based strain sensors have a good future because of its applications in many fields including human diagnosis. Since graphene is highly transparent, it can be used in touchscreens as well. It is expected that the market for strain sensor based on graphene can be as high as 4.5 billion US dollars.

6.1.12 Graphene for Electric Power Generation

6.1.12.1 Fuel Cells

There are various types of fuel cells, but those working at room temperature and using non-reactive gases, i.e., those gases that will not degrade graphene, can be considered to be used in developing electrode for fuel cells. For example, alkaline fuel cells consists of two electrodes (anode) for the oxidation of fuel like hydrogen, and a cathode for the reduction of fuel like oxygen. These two electrodes are separated by a membrane soaked with solution like 30% KOH solution. The anode has to be porous and a conductor so that hydrogen gas can diffuse into the pores, which is also soaked with alkali like KOH. At the electrode, three body reactions takes place. Oxidizing hydrogen to hydrogen ion and electron. The electron is transported to cathode by the external circuit. Hydrogen ion diffuses into the electrolyte to reach near the cathode. At the cathode-like anode, the porous electrode allows oxygen to diffuse into its pores where in the presence of alkali it can be reduced by the electrons coming to the cathode via the external circuit. In this process, hydroxyl ions are produced (OH^-). These hydroxyl ions combine with hydrogen ions (H^+), producing water. This is the entire process occurring at these two electrodes. The problem associated with such a process is that the electrolyte should not be allowed to diffuse out of the electrode to reach the external side of the electrode. If this happens, it will short circuit the electrode. The pores of the electrode should allow the electrolyte to get soaked, but prevent it to diffuse out of the electrode. At the same time, gas should be allowed to reach the pores to carry out the oxidation.

Reduction processes: This electrochemical process is assisted by the type of electrocatalyst loaded on the electrode (i.e., pores of the electrode loaded with electrocatalysts). These conditions can be met by making the pores of the electrode hydrophobic, while at the same time the hydrophobicity of the electrode should not prevent the electron transfer process. This is the place graphene can play a very useful role. Graphene is hydrophobic, a very good conductor and can allow gas to diffuse through its surface. Graphene can be loaded with suitable catalysts like platinum, nickel, etc. Multilayer graphene can play a very useful role in such electrochemical processes. Scientists are working on this topic to make fuel cells either working in an alkaline solution or in an acidic solution.

6.1.12.2 Solar Cells

Similarly in photovoltaic cell, there is a need to develop a semiconductor (n and p-type) with a band gap of 1.4 eV (if it is to be used under solar radiation). The semiconductor should be a good conductor for electron and holes and its Fermi energy should be such that it can form a good depletion region. Graphene has a zero band gap, which is not suitable for photovoltaic cells unless its band gap could be increased to about 1.4 eV. Graphene consists of 100% sp^2 carbon. Some of the sp^2 carbon be converted into sp^3 carbon so that a controlled amount of sp^3 could be introduced into graphene. This is because graphene with zero band gap has 100% sp^2 and diamond with 100% sp^3 has band gap 5.5 eV. Therefore, if possible conditions are developed so that the concentration of sp^2 carbon is decreased and a controlled amount of sp^3 carbon is added into the system, there is a possibility of opening the band gap of graphene to the desired values. Sharon *et al.* have shown the possibility of such conversion and that band gap changes by changing the ratio of sp^2 and sp^3. If such an alteration can be done with graphene, then could have applications in photovoltaic cells.

For making a Shottky type cell or organic cell, graphene can be used as a metal provided its work function matches suitably with the conjugate semiconductor for making Schottky type cell or matches with HUMO level of dye to be used for developing organic cells.

6.1.13 Graphene as a Compliant Substrate

Scientists have observed that the reactivity of monolayer graphene largely depends upon the type of substrate used for its deposition. This behavior is very strange and has not been observed with any other material. It has been observed that graphene deposited on a smooth pure surface of even boron nitride (hexagonal) is unreactive. The advantage of this behavior is that where stable graphene is required, it can be deposited over boron nitride rather than on SiO_2. While this behavior is useful in one way, conventionally prepared graphene over a substrate may not show the same behavior though the process of deposition may be the same. Some may show semiconducting behavior and some may show metallic behavior. This is more important when large numbers of samples are to be prepared. With careful manipulation of substrate, one could tune the type of graphene required to show either metallic or semiconducting behavior.

Since this special behavior is observed with monolayer graphene, it would suggest that this property is controlled by the interaction of π-electron rather than σ-electron. Moreover, when graphene shows metallic behavior, it would suggest that π-electrons are unaffected by the surface of the substrate. However, if π-electrons undergo some kind of interaction with the surface of the substrate, it will decrease the freedom of π-electrons to move on the surface, resulting in semiconducting behavior. This means that the collision of electrons is not affected by temperature, and hence shows semiconducting behavior. This proposed mechanism, however, needs to be experimentally confirmed.

6.1.14 Graphene as Template for New Materials

A template is formed by depositing two different types of material one on top of other, such that they form like a single-layer crystalline material. Such types of materials has many engineering applications. Graphene can find many engineering applications by integrating two-dimensional templates of graphene with other materials such that both have the same crystalline structure. One example is the deposition of boron nitride over graphene forming a single layer of one atom thick. Another example is growing graphene over copper by CVD process. The chemical vapor deposition of MoS_2 over the graphene surface by CVD process is another example. Single crystalline hexagonal flakes of MoS_2 of size up to micrometers can be grown over graphene. A MoS_2/graphene template can be used for the development of optical and electronic devices.

6.1.15 Biodevices Based on Graphene's Chemical Properties

Apart from a large surface area, atomic thickness, the modifiable chemistry and molecularly obtainable structure are the major properties that will allow the possible use of antibody-functionalized graphene sheets for biodevices that can be used for detection and diagnostic purposes in mammalian and microbial systems (Mohanty *et al.* 2008). The use of graphene is being envisaged for Single-Molecule DNA Sequencing Technologies also (Xu *et al.* 2009).

Graphene will pave the way for bionic devices in living tissues that could be connected directly to neurons of the injured spine and could help in reusing limbs.

For substantially influencing graphene-based devices, graphene needs chemical modification of its basal plane or its edges.

6.1.16 Graphene in Healthcare

6.1.16.1 Cytotoxicity a Concern

There has been tremendous progress in use of graphene for various biomedical applications, e.g., drug delivery, cancer therapies and biosensing.

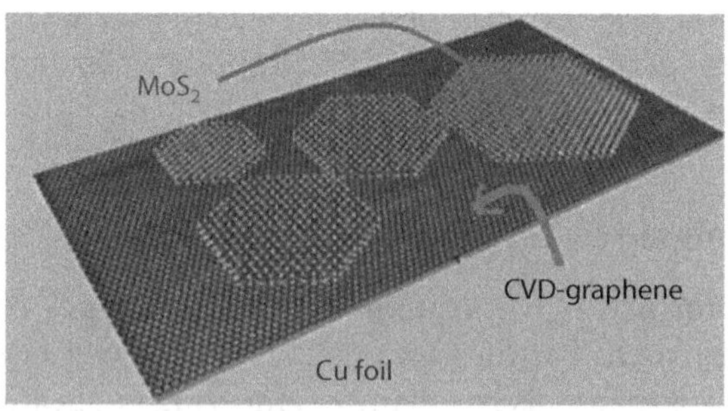

Figure 6.4 van der Waals Epitaxy of MoS2 layers using graphene as growth templates. Copyright© 2012 American Chemical Society; Courtesy: Shi *et al.*; *Nano Lett.*, 2012, 12 (6), pp 2784–2791.

At the moment, though most of the findings are promising, but they are but at a preliminary stage. However, apart from the opportunities, this emerging field faces many challenges also because there are many unresolved challenges in graphene-based nanomedicine. One of the major concerns in using graphene for drug delivery has been the consideration of its short and long term toxicity. The toxicity of graphene is closely associated to its surface functionalization. But nanographene with biocompatible coatings, e.g., with PEG, exhibits a very less toxicity both *in vitro* and *in vivo* and shows excellent stability in the presence of high-concentration salts and proteins. PEG-coated graphene even at 100 mg/l concentration shows negligible *in vitro* toxicity to HCT-116, OVCAR-3, U87MG, MDA-MB-435 and MCF-7 cell lines. However, why only nanographene affects the *in vivo* behaviors is a matter of investigation.

Cytotoxicity of Graphene—Some of the efforts regarding the toxicity of graphene has shown that graphene synthesized by the CVD technique increased the activation of caspase 3 an (apoptosis marker, released the lactate dehydrogenase and generation of reactive oxygen species (ROS), in neural pheochromocytoma-derived PC12 cells (Zhang *et al. 2010*).

Cytotoxicity of Graphene Oxide (GO)—The GO above 50 mg/l concentration was found to induce cytotoxicity of human fibroblast cells (Wang *et al.* 2011), and at concentration up to 85 mg/l only lightly decreased A549 cells proliferation rates but did not induce apoptosis (Hu *et al.* 2010). Moreover, intravenous injection of GO into rats during *vivo* toxicity study has shown that it mostly accumulates in lungs and induced dose-dependent (obvious toxicity at a dose of 10 mg/kg pulmonary toxicity). This could be because GO without further surface functionalization is not stable in physiological environments. Trials of intravenous injection of radio-labeled graphene oxide + PEG of 10 to 50 nm for photothermal treatment did not show obvious toxic side effects in mice in 40 days, which was mainly localized in the reticuloendothelial system, including the liver and spleen, with negligible accumulation in the lungs. Moreover, a dose of 20 mg/kg over a course of three months was gradually excreted from mice without causing any toxicity. For this purpose an amine-terminated, branched PEG was used to functionalize GO.

Cytotoxicity of Reduced Graphene Oxide (rGO)—According to Hu *et al.* (2010), rGO has shown very high cytotoxicity when assessed by treating A549 cells with Hydrazine hydrate (a reducing agent) whereas GO did not show much cytotoxicity.

6.1.16.2 Graphene for Drug Delivery

Almost four years after the discovery of graphene, bioscientists started paying attention to study the potential of applications of mostly nanographene in bio-medicines, e.g., as a cancer drug carrier. The first reported use of graphene in drug delivery came to light in 2008 by Dai's group (Liu *et al.* 2008 and Sun *et al.* 2008). Graphene was easily functionalized with desired molecules and coated with biocompatible amine terminated branches polyethylene glycol (PEG). It was found to be suitable for high passive *in vivo* tumor uptake and photothermal ablation of tumor. Very high loading of anticancer drugs SN38 and Doxorubicin have been effectively achieved on the graphene surface via ϖ-ϖ stacking for intracellular drug delivery. The high drug loading efficiency was due to the extremely large surface area of graphene, where every atom is exposed on its surface.

In an *in vivo* study using several different xenograft tumor mouse models, very high passive uptake of graphene in tumors is reported. Since the intrinsic photoluminescence of nanographene is too weak for *in vivo* imaging, the study was conducted using nanographene labeled by a NIR fluorescence dye for *in vivo* fluorescence imaging.

Though the graphene-based nano-medicine is still in its infancy, it is going to bring unique and novel opportunities in the area of cancer treatment. Scientists are realizing that for the applications of graphene-based drug delivery, there should be approaches towards combining drug delivery with other therapies like photothermal and gene therapy.

6.1.17 Graphene in Textiles and Fabrics

Scientists are trying to make yarn from graphene and use them where electrical properties are needed, for example, making supercapacitors, fuel cells, etc. Whenever there is a need to develop electrically active electrodes, as one uses in making supercapacitors or gas diffusion electrode for fuel cells, graphene yarn is prepared such that its electrical and surface area are not lost. This yarn made from graphene acts like electrically conducting cloth and if the weaving is nicely done, then it could suffice the need of gas diffusion electrode. It has been observed that yarn spun from graphene oxide has the highest capacitance as compared to other carbon materials. In addition, such yarn is very strong, flexible and lightweight.

6.2 Summary

In this chapter, applications of graphene have been discussed. Care is taken to categorize the applications based on graphene's specific properties. Graphene possesses high Young's modulus strength and high stress, and therefore stress is on the applications of graphene where such properties are necessary. Graphene forms composites with polymers very nicely, and hence applications where composites can be applied have been discussed. Graphene forms transparent thin film. There are many places where there is a need for material that is transparent, and yet a good conductor of electricity, like ITO. Graphene can find its place in these areas. In fact, considering its high temperature stability, it can be preferred over ITO film. Graphene possesses high surface area and is a good electrical conductor and many surface states. These are ideal properties to develop biosensors. In fact graphene is preferred as a biosensor material. It has been possible to open the band gap of graphene by suitable substitution. This aspect of graphene has been useful in developing flexible organic solar cella and metal Schottky junctions. Graphene's high thermal conductivity and high electrical conductivity have made it a useful material for developing electronic circuits.

7

Towards Mass Production of Graphene: Lab to Industry (Scaling Up)

A careful analysis of the process of observation in atomic physics has shown that the sub-atomic particles have no meaning as isolated entities, but can only be understood as interconnections between the preparations of an experimental and the subsequent measurements.

Erwin Schrödinger

With the advent of many applications of graphene, there is a need for large-scale production. However, in the present day scenario, it can be said that the type of technique available for the production of graphene still suffers from the problem of mass-scale production of high quality graphene. To prepare perfect graphene film without impurities at a large scale is critically difficult. However, in recent years, there have been many approaches in the mass production of graphene. It is imperative to have an efficient method to produce graphene at a large scale, because needless to say, a graphene market will be driven by the mass-scale production of high quality graphene. As we have seen in chapters 5 and 6, for various specific applications of graphene, there has been a need for differently engineered graphene. Basically, the synthesis methods of graphene can be categorized as top-down and bottom-up approaches.

The top-down approach involves detachment of the graphene layer from already existing structures in the form of graphite. Graphite is composed of many layers of graphene. The techniques used are Micro-mechanical Cleavage, Chemical Exfoliation through Oxidation of Graphite, Electrolytic Exfoliation and Thermal Exfoliation and Reduction.

The bottom-up approach is depositing graphene layer on a substrate using hydrocarbon as the source of carbon. This approach encompasses Chemical Vapor Deposition, Chemical Conversion of CO_2, Unzipping of Carbon Nano Tube, Arc Discharge, Self Assembly of Surfactant, Epitaxial Growth on SiC and Solvo-Thermal methods.

The successful lab-scale production methods of graphene, the possibility of their scaling up and successfully scaled-up techniques using these two approaches, along with the challenges in scaling up are discussed here.

7.1 Exfoliation of Graphite: A Top-Down Approach

The first successful attempt of separating graphene was done by mechanical exfoliation method, followed by many thermal, electrolytic and solvo-thermal methods. Graphene sheets of up to 100 μm in size have been obtained by repeated peeling of three-dimensional graphite crystals.

7.1.1 Micro-Mechanical Exfoliation or Repeated Peeling of Graphite

In micro-mechanical exfoliation, commonly known as the Scotch-tape method, first developed by Novoselov and Geim in 2004, adhesive tape is used to detach graphene from graphite crystals. In the first peel, many layers of graphene get detached from graphite. To get mono, bi or a few layers of graphene, the graphene is peeled repeatedly from the adhesive tape. Then finally the tape is attached to a substrate and glue is removed by dissolving it in solvent such as acetone.

The procedure: Followed by Novoselov *et al.* (2004) involves an initial preparation of 5 μm-deep squares of various sizes from 20 mm to 2 mm of HOPG (Highly Oriented Pyrolytic Graphite). On a glass substrate a 1 mm thick layer of photoresist is spread by spin coating. The pieces of HOPG are mounted on the photoresist layer and heated or baked so as to allow the attachment of HOPG blocks on the glass.

Prepared HOPG samples are then repeatedly peeled off using adhesive tape and finally the thin flakes left in the photoresist are released in acetone, as acetone dissolves the photoresist of glue.

Silicon wafers (SiO_2 with thickness = 300 nm) are used as a substrate to load graphene. Prior to their use, silicon wafers are thoroughly washed with water and then cleaned by being dipping in propanol and ultra-sonicated. The cleaned Si wafers are dipped in the solution. Graphene flakes get captured on the surface of wafer. To remove any thick flakes attached to the wafer, they are ultra sonicated in propanol. The thin flakes remain attached to SiO_2, due to van der Waals and/or capillary forces.

There has been some improvement by different scientists of this exfoliation procedure to suit particular requirement. However, the initial process followed by Novoselov *et al.* still remains the standard procedure. By this method, good quality graphene from single layer to a few layers has been successfully exfoliated.

Challenges in scaling-up micro-mechanical exfoliation method: Though the quality of graphene prepared in this way is very high with almost no defects, there is no controllability of the size and thickness of graphene as it varies considerably (from few nm to µm) as well as person to person.

7.1.2 Liquid Phase Chemical Exfoliation of Graphite

Chemical exfoliation of graphite is done by dispersing graphite in an organic solvent (0.3 mg/mL sodium cholate) with almost the same surface energy as graphite, so that the energy barrier is removed to facilitate the detachment of graphene layer (Lotya *et al.* 2010). This solution is then either ultra-sonicated for 100–400 hrs or voltage is applied to the solvent to split the graphite into individual graphene sheets as suspension. Then the solution is centrifuged between 500 and 2000 rpm to remove the thicker flakes. By this process, graphene flakes of 1–10 stacked monolayer with up to 20% of flakes containing just one layer is obtained. The average flake consists of ~4 stacked monolayers and has width of ~ 1µm to ~400 nm. The mean flake length of this graphene varies from ~1µm to ~500 nm. It must be mentioned here that prolonged centrifugation reduces the length of graphene. To get relatively defect-free graphene it is imperative to optimize the centrifugation rate, or at least it should be kept <2000rpm.

Challenges in scaling-up chemical exfoliation of graphite: This method yields even better quality of graphene that is obtained by micro-mechanical exfoliation, but the size is very small and there is a lack of controllability also. However, it is one of the most powerful methods for mass production of graphene. By this method, a very large quantity of very small graphene can be obtained.

7.1.3 Liquid Phase Aqueous Exfoliation of Graphite Oxide

Another approach is to first oxidize the graphite pellets by chemical treatments.

The procedure: The standard procedure used for oxidation of graphite is by reacting graphite with strong oxidants such as $KMnO_4$, $KClO_3$, and $NaNO_2$ in the presence of nitric acid or in the presence of mixture of nitric and sulfuric acid (Hummers method). This process yields GO composed of graphene oxide sheets stacked with an interlayer spacing between 6 and 10 Å depending on the water content. Graphite oxide has several functional groups such as hydroxyl or epoxide. Hence, it is water soluble. Therefore, the principle of liquid phase exfoliation is applied by sonication. During sonication the layers become negatively charged and recombination is inhibited by the electrical repulsion. By oxidizing graphene flakes and then floating them in water, the graphene flakes form a single sheet and bond very powerfully, which is called graphene oxide paper. Graphene oxide paper has found applications in membranes with controlled permeability, supercapacitors for energy applications and graphene electronics for communications and imaging technologies that require ultrafast transistors.

The next step is to reduce the GO. Reduction of graphene oxide is necessary to restore electrical conductivity, because GO is electrically insulating and thermally unstable. This is done after centrifugal separation of graphene oxide nanosheets; they are reduced *in situ* to the original graphene state, so as to achieve high electrical conductivity. For the reduction of GO sheet, either microwave irradiation or strong chemical reductants such as hydrazine (N_2H_4), sodium boronhydride ($NaBH_4$) or chlorogenic acid (Choi *et al.* 2013) are used.

This is one of the most promising methods for large-scale production of graphene.

Challenges in scaling-up aqueous exfoliation of graphite oxide: Are that the complexities of this method are higher and quality is also not very high.

7.1.4 Thermal Aqueous Phase Exfoliation of Graphite Oxide

An improvisation of the aqueous-based graphene oxide for easy scaling up is done by thermal treatment for exfoliation (Schniepp *et al.* 2006), which simultaneously reduces GO also. In this process, GO is dispersed in aqueous solutions with surfactant and sonicated for a prolonged period along with giving thermal shock procedure to simultaneously reducing the GO and exfoliating the graphite in monolayer graphene flakes, which is further enriched by centrifugation and then graphene is deposited as a thin film on a substrate.

The procedure: This procedure involves the treatment of GO with oxidizing solution of sulfuric acid, nitric acid, and potassium chlorate for >96 h under very rapid heating. Complete oxidation is confirmed by monitoring the disappearance of the 0.34-nm inter-graphene spacing. The new graphene exhibits 0.65–0.75 nm inter-graphene spacing. Formation of GO is facilitated due to thermal expansion of graphite and the presence of oxidizing agents. This GO is dried and charged into a quartz tube and purged with argon. Then it is rapidly heated (>2000°C/min) to 1050°C. The thermal exfoliation of GO into individual sheets occurs through the evolution of $CO_{2.}$ However, though this process preserves many of the properties of single-layer graphene, it yields some multilayer graphene also.

Another improvisation that is similar to oxidation is done by intercalating the graphite with small molecules, using a thermal or plasma process. This process helps in delaminating into single sheets.

Challenges in scaling-up thermal aqueous phase exfoliation of graphite oxide: This process requires a long time, high temperature and fast heating rate. It is also difficult to decide when to stop the experiment because it requires a stopping reaction when spacing between the layers is 0.65–0.75nm.

7.2 Length-Wise Unzipping of Carbon Nanotubes (CNT)

From a topological perspective, a carbon nanotube (CNT) can be considered as a graphene sheet rolled into a tubular form as a single-wall carbon nanotube (SWCNT) or many graphene sheets rolled into multi-wall carbon nanotubes (MWCNT), having concentric rings on many walls. The unzipping of the CNT method is more suitable for producing suspension of well defined graphene nanoribbons (GNR). Unzipping of CNT produces multilayer GNRs with various widths and lengths of up to several microns. Most of the application-based research, e.g., for graphene-based paints and inks, printed electronics, electromagnetic shielding, barrier coatings, heat dissipation, supercapacitors, smart windows, etc., is being completed using GNR. It can be said that CNT is facing stiff competition from its two-dimensional cousin, graphene. No wonder CNT is being unzipped into graphene. Length-wise splitting or unzipping of both SWCNT and MWCNT has been achieved by different methods, including:

1. Selective etching or plasma etching method
2. Oxidizing method
3. Alkali-metal atom insertion method
4. Transition metal (nano particles) catalyzed unzipping
5. Hydrothermal cethod
6. Sono-chemical method

7.2.1 Selective Etching or Plasma Etching Method

A standard procedure for producing GNRs having smooth edges and a narrow width distribution (10–20 nm) by unzipping MWCNT by plasma etching of nanotubes partly embedded in a polymer (Figure 7.1) film was developed by Jiao *et al.* (2009).

This procedure requires MWCNT dispersed in aqueous solution. Dispersion is done by sonicating it in the presence of 1% Tween 20 and then centrifuging it to remove MWCNT aggregates (if any formed).

A silicon substrate pre-treated with aqueous solution of 3-amino-propyl-tri-ethoxy-silane (APTES), rinsed with water and then blowdried is used as a substrate to deposit MWCNT suspension. Any remaining Tween 20 is removed from the sample by calcining it at 350°C.

Onto MWCNT, PMAA is spin-coated and baked on a hot plate at ~170°C. After cooling the PMMA–MWCNT film is peeled off in KOH solution at 80°C and then washed with water. This film is then deposited on a Si substrate heated at 80°C.

To etch away the PMAA–MWCNT film, ten-watt Ar plasma is applied at the base pressure of 40 mTorr, and PMAA film is then lifted using water and adhered on an APTES treated SiO_2/Si substrate with a prefabricated Pt/W marker array. PMAA is then removed by acetone treatment and calcinations. These narrow graphene ribbons have nearly atomically smooth edges with high conductance of up to 5e2/h.

Challenges in scaling-up unzipping of CNT by selective etching or plasma etching method: This method is very sensitive based on the time

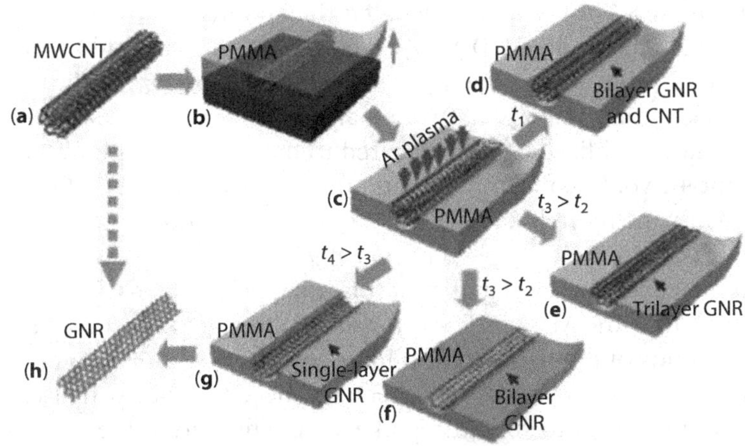

Figure 7.1 Making graphene nanoribbon (GNR) from CNTs (a) A pristine MWCNT used as the starting material (b) The MWCNT deposited on a silicon substrate and then coated with a film of PMAA (c) The PMAA-MWCNT film was peeled from the Si substrate, turned over and then exposed to an Ar plasma d–g, several possible products were generated after etching for different times; GNRs with CNT cores were obtained after etching for a short time t_1 (d) tri-, bi- and single layer GNRs were obtained after etching for times t_2, t_3 and t_4, respectively ($t_4 > t_3 > t_2 > t_1$; e-g). (h) The PMAA was removed to release the GNR. Copyright © 2009, Rights Managed by Nature Publishing Group; Courtesy: Jiao *et al.* 2009, *Nature*, Vol. 458, No.16, p.878, 2009.

PMMA deposited over graphene is exposed to argon plasma. Depending upon the time of exposure under the Argon plasma, it can give mono-, bi- and multiple layer graphene. Thus it may be difficult to get graphene of the desired number of layers. This technique may be useful if multilayer graphene is required.

7.2.2 Oxidizing Method

Almost at the same time Kosynkin *et al.* (2009) have presented another method to synthesize single layer nanoribbons of graphene by cutting MWCNT lengthwise. It is a solution-based oxidative process that yields nearly 100% GNR with high water solubility and restored electrical conductivity. Potassium permanganate and sulphuric are used as oxidizing agents. The H_2SO_4 helps in exfoliating the MWCNT. The procedure follows suspension of MWCNTs in concentrated sulphuric acid (H_2SO_4) for about an hour followed by treatment with 500 wt% potassium permanganate ($KMnO_4$) initially at room temperature and then heated to 55–70°C for an hour till all the $KMnO_4$ is consumed. It is then poured onto a mixture of ice and hydrogen peroxide (H_2O_2). The solution is then removed by filtration over a PTFE (polytetrafluoroethylene) membrane. The remaining solid is first washed with acidic water, then ethanol/ether.

The isolated oxidized GNR is then reduced by treating with water + SDS surfactant +1 vol % concentrated ammonium hydroxide (NH_4OH) and 1 vol 1% hydrazine monohydrate ($N_2H_4.H_2O$). The solution is covered with a thin layer of silicon oil and then heated at 95°C.

To understand the mechanism of this unzipping process, it is advisable to examine how the MWCNT was formed. When graphene folds to make CNT, two ends of graphene fold to form a cylindrical shape, and carbons present on the two ends of the graphene will need to modify their configuration from pure sp^2 to something like sp^3 configuration (i.e, it is neither pure sp^2 nor pure sp^3 configuration). It is possible that these carbon atoms may undergo chemical oxidation more favorably than other carbon atoms (i.e., pure sp^2 carbon). Thus one can visualise the opening of CNT in one straight line as shown in Figure 7.2. These carbons after opening may form quinone like configuration as shown in Figure 7.2. If this process occurs then unzipped graphene oxide should be in a perfect rectangular shape.

Challenges in scaling-up unzipping of CNT by oxidizing method: Though permanganate treatment is a scalable process, it produces highly oxidized GNRs that are just like Graphene oxide (GO). These oxygen-containing functional groups present on the basal plane of the oxidised GNR

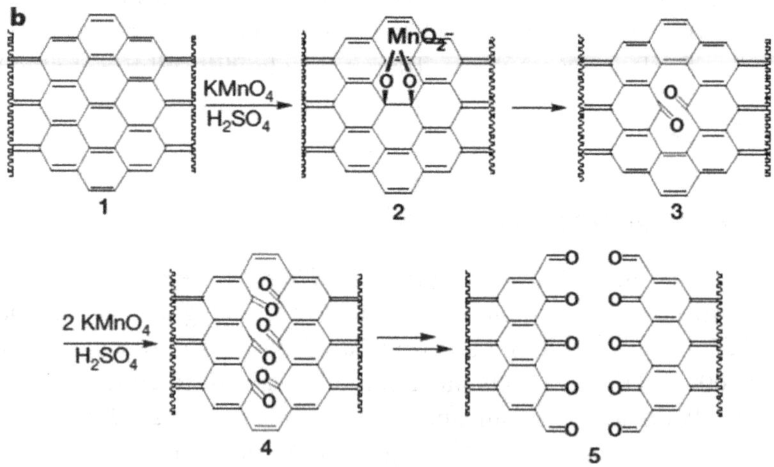

Figure 7.2 Exhibiting the mechanism of unzipping CNT to form GNR. Copyright © 2009, Rights Managed by Nature Publishing Group; Courtesy:Kosynkin *et al.*; *Nature*, 2009, 458 (7240)

can be removed by annealing in hydrogen at 900°C, but always some fraction of defects remain, which is not very desirable because it lowers the electrical conductance dominated by a Mott variable range hopping mechanism instead of the desirable ballistic transport (Sinitskii *et al.* 2009).

7.2.3 Alkali-Metal Atom Insertion Method

Alkali metals are inserted between the concentric walls of MWCNT. This causes graphene layers to peel off. Kosynkin *et al.* (2011) have prepared GNR free of oxidized surfaces with a 100% yield by splitting MWCNT with potassium vapor.

The methodology involves the use of MWCNT without any treatment. MWCNT and potassium were taken in 1:3 ratios in a glass ampoule and evacuated and sealed. This is a hazardous and tricky step because potassium is highly reactive. Therefore, care and precautions are to be taken, i.e., it should be performed in a fume hood with face shield. Potassium was melted over MWCNT under vacuum (0.05 Torr). An ampoule was placed in a furnace and heated to 250°C for 14 h (Figure 7.3). The ampoule becomes golden-bronze colored. The ampoule now contains potassium intercalation compound and some unreacted metal as silvery droplets. It is cooled to room temperature and then opened either in a dry box or in a nitrogen-filled glove box, and ethanol is slowly added to it. When ethanol is added the mixture bubbles and much of the heat release is dissipated by the released gas (hydrogen). The product, i.e., longitudinally split MWCNTs

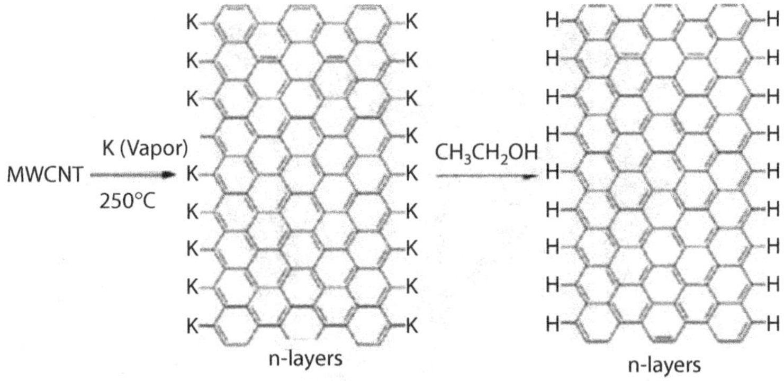

Figure 7.3 Chemical schematic of the splitting processes where ethanol is used to quench the aryl potassium edges; only a single layer is shown for clarity, while the actual number of GNR layers correlates with the number of concentric tubes in the MWCNT. Taken from Wei Lu 2012, Rice University Ph. D. Thesis.

as a black, fibrillar powder (Figure 7.4); is then removed from the nitrogen enclosure and collected on a PTFE membrane (0.45 μm), washed with ethanol, water, ether and finally dried in vacuum.

To exfoliate potassium split layers of MWCNTs, they are dispersed in chlorosulfonic acid and ultrasonicated. Chlorosulfonic acid is another hazardous material in this method that is to be handled carefully because it is corrosive and reacts violently with water to form HCl and H_2SO_4.The mixture is quenched by pouring ice on it. The suspension containing separated GNR is then filtered through a PTFE membrane. The filtered black material is dried under vacuum and dispersed in dimethylformamide (DMF) and sonicated.

Challenges in scaling-up unzipping of CNT by alkali-metal atom insertion method: This process, though simple, requires handling potassium metal along with MWCNT under vacuum and heating it at 240°C for 14 h. The preparation of a small quantity may be possible by this method,

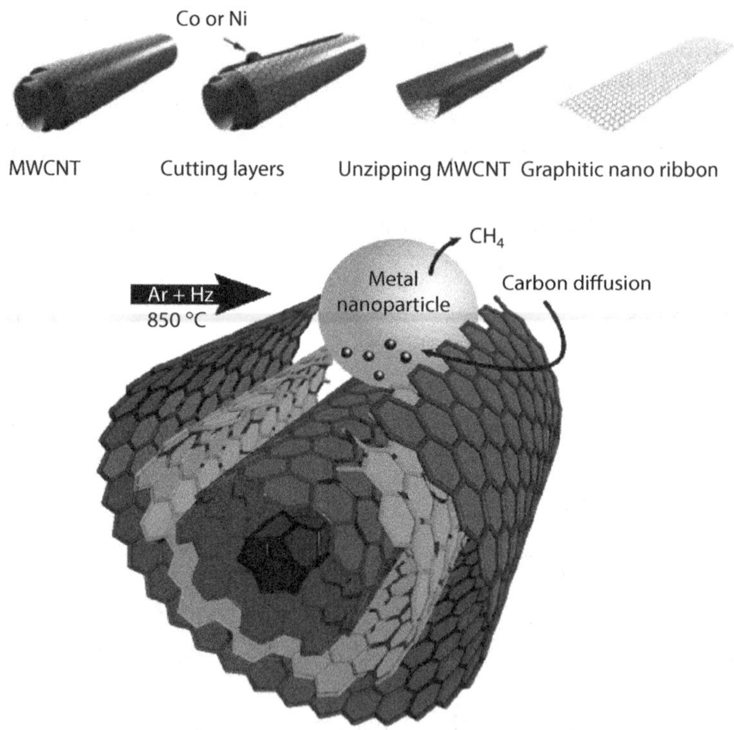

Figure 7.4 Longitudinal cutting of MWCNT to form graphitic nanoribbons using metal clusters as nanoscalpels. Copyright© American Chemical Society; Courtesy: Elias *et al.* *Nano Lett.* 2010, 10, 366.

but making large quantity of graphene oxide by this process may be not advisable. The mechanism of the formation is more or less similar to the previous method (7.2.2).

7.2.4 Catalytic Unzipping of Carbon Nanotubes

In this method, nano Palladium (Pd) is used as a suitable catalyst. Pd is taken in an oxygen-containing liquid medium and microwave irradiated. This process cuts open the SWCNT, DWCNT and MWCNT (Janowska *et al.*, 2009). Palladium acts as a scissor to cut CNT lengthwise (Figure 7.5). Using force field studies, it has been demonstrated that the presence of nanocatalysts Pd and oxygen next to vacancies facilitate the unzipping process to form graphene from CNT by significantly lowering the corresponding energy barrier.

Challenges in scaling-up unzipping of CNT by catalysts: The insertion of catalytic metal nano particles gives a rather low yield of GNRs. Moreover, the GNRs are usually consisting of 50–100 layers (Elías *et al.* 2010).

7.2.5 Hydrothermal Method

Recently Shinde *et al.* (2014) have developed a counter ion–dependent method for unzipping of MWCNT (Figure 7.6). By this method 80% highly conductive and transparent GNRs can be obtained. MWCNT dispersed in water is used for unzipping. To this aqueous dispersed MWCNT, several counter ions from K_2SO_4, KNO_3, KOH or H_2SO_4 are added separately and ultra-sonicated. During ultra-sonication, selective intercalation of K ions occur. Simultaneously sulphate and nitrate ions also act as a co-intercalant along with potassium ions. This causes exfoliation of multi-walled carbon nanotubes. The suspension is transferred to

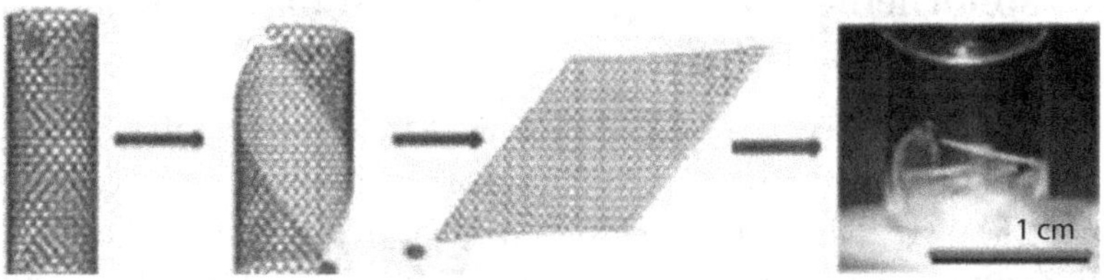

Figure 7.5 Schematics of catalytic unzipping of carbon nanotubes in the presence of Pd nanoparticles, as a pair of scissors to cut nanotubes lengthways and an oxygen-containing liquid medium by microwaves irradiation. Courtesy: Janowska et al. *Applied Catalysis A: General* Volume 371, Issues 1–2, Pages 22–30, 2009.

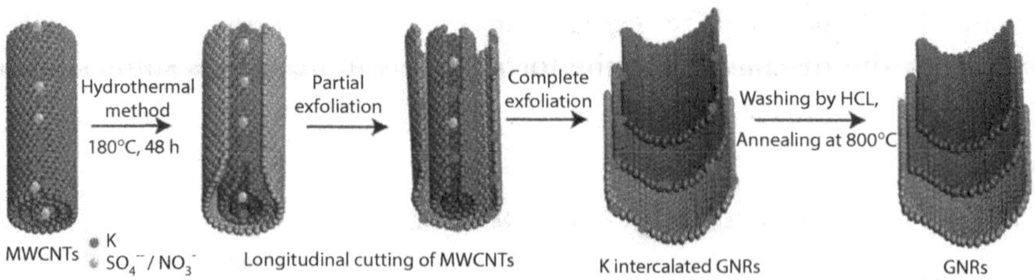

Figure 7.6 Schematic representation of the sequential longitudinal cutting of MWCNTs using potassium and sulphate/nitrate intercalation by hydrothermal method to generate highly transparent GNRs. Courtesy: Copyright © 2014, Rights Managed by Nature Publishing Group Dhanraj B Shinde *et al.* 2014 *Scientific Reports* 4, Article number: 4363 doi:10.1038/srep04363.

a poly tetra fluro-ethylene (PTFE) lined autoclave (50 ml) and annealed at 180°C to complete the process. The suspension is then cooled to room temperature. The black suspension (containing GNR) thus obtained is washed with dilute HCl and distilled water (till it reaches a neutral pH) to remove excess potassium from GNRs.

Challenges in scaling-up unzipping of CNT by hydrothermal method: The reasons for the use of various salts of potassium being used for unzipping CNT is not clear. Potassium being reactive, its role can be understood, but anions are not expected to interact with carbon atoms. Perhaps the energy involved in hydrothermal process may be suitable to help potassium ions to interact with carbon atoms. Nevertheless, working with salts of potassium is less dangerous as compared to working with potassium metal. This process can be used to prepare a large quantity of graphene oxide. But the role of different salts of potassium needs to be examined.

7.2.6 Sonochemical Unzipping of Multi Wall Carbon Nanotubes (MWNTs)

For sonochemical unzipping, the pristine MWCNT (synthesized by arc discharge) are calcined in air at 500°C for an hour, so as to remove any impurity. MWCNT is then etched/oxidized at defect sites and ends, without oxidizing pristine sidewalls of MWCNTs. The MWCNT and PmPV [poly (m-phenylenevinylene-co-2, 5-dioctoxy-pphenylenevinylene] are then dispersed in DCE (1, 2-dichloroethane) and sonicated to unzip the MWCNTs into GNRs (Jiao *et al.* 2009). The remaining CNTs after sonication were removed by ultra-centrifugation (40,000 rpm for 2 h). This process yields >60% GNR. The remaining CNTs can be used for repeated unzipping again. GNR suspension is then coated on SiO_2/Si substrates and calcined in air at 350°C for cleaning.

Challenges in scaling-up unzipping of CNT by sonochemical method: This is a preferred method for GNR only. Though compared to chemical exfoliation it is a more expensive method, it has already been scaled up and is produced and marketed by the ton. By the CNT unzipping method, dimensions of the GNR can be predicted by the initial CNT length and diameter.

7.3 Chemical Vapor Deposition (CVD) Method

CVD is a bottom-up approach for the direct synthesis of graphene from carbon sources or precursors such as methane, which is then transferred to a substrate. By CVD method large-area, single and few-layer graphene sheets are grown on metal foil substrates, for example, Cu foil (Bae *et al.* 2010). The graphene sheet deposited on the Cu foil is then transferred to a dielectric surface or other substrate of interest. By this method, square meters of monolayer graphene has been fabricated (Li *et al.* 2009) and transferred to a 200 mm Si wafer. It shows great promise for its use in touchscreens. CVD has the potential and promise to become a commercial manufacturable technology.

CVD process involves deposition of carbon film having graphene morphology from the vapor phase of carbon. A typical CVD setup has three main systems: (i) a gas carrying system (ii) a tubular furnace, which is the reactor surrounded by a heater and a thermocouple is placed to monitor the temperature (iii) a gas removal system. Through the gas carrier system, the gas is delivered into the reactor. The flow of gas and the mixing of desired gases are controlled by Mass Flow Controllers prior to their entry into the heating zone of the reactor, where the chemical reaction takes place and vapor is deposited as solid on a substrate. Transport kinetics of gas is an important critical parameter for synthesis of graphene. The unreacted gases and by-products are removed by the gas outlet. System pumps or valves are attached to the outlet for adjusting the pressure or creating the vacuum. In a nutshell the carbon vapor is generated from hydrocarbon sources such as methane and then passed to the furnace where the reaction takes place (i.e., conversion of carbon vapor to graphene) and graphene is deposited on the metal substrate placed in the hot zone of the furnace. The substrate commonly used is Cu foil, which is first annealed by H_2 and Ar so as to increase the Cu grains.

Cleaning of copper is one of the important parameters for getting a desired type of graphene sheet. For fabricating good quality it is imperative that the cooling of the furnace is done rapidly under Ar gas.

A mechanism of graphene fabrication: On Cu foil depends on the low solubility of carbon in Cu, diffusion of C atoms on Cu surface, absorption and de-absorption of hydrocarbon on Cu. A pictorial mechanism of growth mechanism is shown in Figure 7.7 and Figure 7.8. Commonly used precursor is methane. It is a relatively stable hydrocarbon compound due to strong C-H bond, as a result, decomposition occurs at temperatures >1200°C. Ethane or acetylene are not used because they rapidly decompose at high temperature.

The hydrocarbon gets decomposed at a high temperature and forms a carbon atom that aggregates on the Cu surface to initiate graphene nucleation centers. Further diffusion and attachment of carbon atoms to nucleation centers helps in continuation of graphene film/sheet formation.

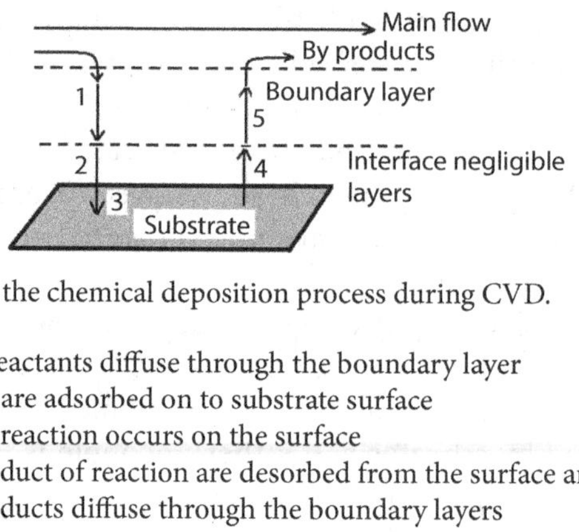

Figure 7.7 Explains the chemical deposition process during CVD.

1. First the reactants diffuse through the boundary layer
2. Reactants are adsorbed on to substrate surface
3. Chemical reaction occurs on the surface
4. The byproduct of reaction are desorbed from the surface and
5. The byproducts diffuse through the boundary layers

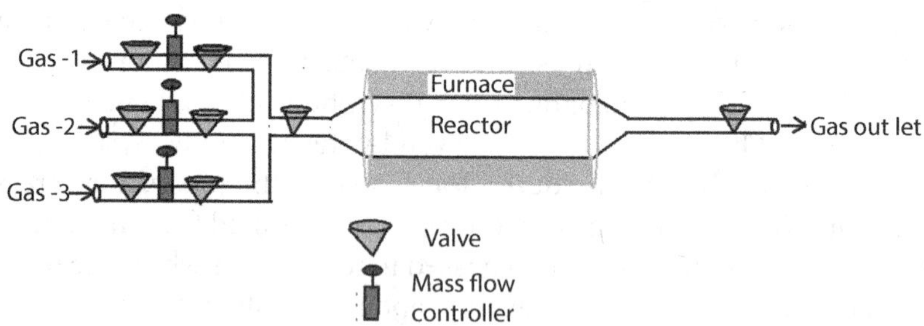

Figure 7.8 A schematic diagram of tubular CVD furnace. Gas flows are regulated by MFCs and fed into the reactor through a gas-distribution unit. Chemical deposition takes place in the heated reactor. The exhaust gases are removed by vacuum pump.

People have tried to use other transition materials as substrate for graphene, e.g., Fe, Co, and Ni. But their capacity to decompose hydrocarbon is very high. Hence, they are not suitable for the fabrication of mono or bilayer graphene, whereas the low decomposition rate of methane on Cu is suitable for controlling the number of graphene layers. However, Cu foil with smooth surface is preferred.

Prior to use, Cu foil is processed by treating it with dilute acid and then ultra-sonicating in acetone and finally annealing at low pressure, which makes the surface smoother. The purity of Cu foil affects the graphene layer formation. According to a report, Cu foil with higher purity (2638–2641 cm^{-1}, I_{2D}/I_G: over 3) produces monolayer and Cu foil with purity (2641–2646 cm^{-1}, I_{2D}/I_G:1.8-2.4) produces bilayer graphene. By CVD methods, few layers of high quality large size (up to cm), defect-free graphene sheets are produced. Electrochemical cleaning also helps to grow graphene with large grain size.

Challenges in fabrication of graphene by CVD method: Challenges encountered in synthesizing graphene by CVD methods are that CVD is an expensive process as it consumes lot of energy and the metal substrate onto which graphene is grown has to be removed, so a method should be developed to recover these metals. Moreover, controlling the domain (grain) size, ripples, doping level and the number of layers, number and relative crystallographic orientation of the graphene layers are required. Finally, the transfer process needs to be improved so as to minimize the damage done to graphene.

CVD produces small amounts of sheets and it is difficult to modify the surface structure by CVD method, therefore, graphene obtained from this method is not suitable for polymer nanocomposites because fabrication of nanocomposites requires a large amount of graphene sheets preferably with modified surface structure.

7.4 Epitaxial growth of Graphene on Silicon Carbide

For various industrial applications, graphene is required to be supported by a substrate. Here we discuss use of SiC as substrate for graphene and how it interacts with the graphene layer and the effect of the interface on graphene atomic and electronic structures.

As the name suggests, Silicon Carbide $(SiC)_n$ is a compound of silicon and carbon. Both are present in equal number. As can be seen in the Figure 7.9, each silicon atom is attached to four carbon atoms and each carbon atom is attached to four Si atoms.

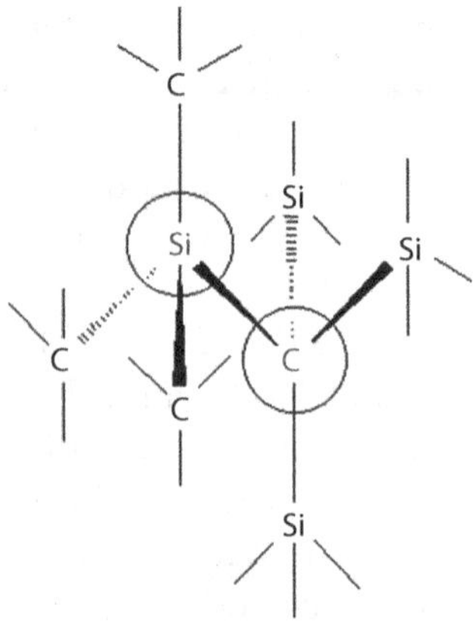

Figure 7.9 Molecular structure of silicon carbide

The layer of epitaxial graphene grown on SiC is weakly coupled to the substrate by van der Waals Forces. The weak van der Waals force provides the cohesion of multilayer graphene stacks also without affecting the electronic properties of the individual layers. Hence, the electronic properties of multilayered epitaxial graphenes are similar to that of a single layer.

As suggested by Geim in a letter to the American Physical Society (2010); J. Van Bommel *et al.* were the first to report the epitaxial growth of graphite in 1974. They had shown that heat treatment of the SiC (0001) crystal surfaces showed a layer of graphite formation because of evaporation of silicon, and thus the formed graphite layer has a distinct crystallographic relation to the SiC crystal. The characterization studies have revealed that it is monocrystalline on the Si-face and mostly polycrystalline on the C-face. In 2004 Walt de Heer from Georgia Tech pioneered the growth of thin graphitic layers using SiC as substrate and reported it to be the material for future integrated circuits.

Though like graphite, SiC also occurs in nature as a rare mineral (as Moissanite), but nowadays SiC is mostly synthesized because it is used for many high-powered electronics. Silicon carbide wafer is used to produce epitaxial graphene by graphitization at high temperatures. At high temperature and ultra-high vacuum (Lin *et al.* 2010; Tzalenchuk *et al.* 2010 and Cai *et al.* 2010) or atmospheric pressure (Hackley *et al.* 2009) Si atoms sublime and remaining carbon forms graphitic layers on either carbon or silicon faces of SiC wafer. Methods have been developed to control the growth of the number of graphene layers (Ohta *et al.* 2006).

Prior to using SiC wafers as a substrate, it is first etched by hydrogen so as to produce an atomically flat surface and then used for low pressure

(~10^{-6} torr) and high temperature (1100^{0}C) graphitization. This produces ultra thin layer epitaxial graphene (EG). Like CVD, this method also yields sheets of defect-free good quality graphene. The dimensions of the EG depends on the size of the wafer. Moreover, the face of the SiC used for fabricating graphene, silicon- or carbon-terminated, influences the thickness, mobility and carrier density of the resulting graphene (Magaud *et al.* 2009).

On the Si face, the C layers are in epitaxy grown on SiC. The substrate surface is passivated by the first C layer—the buffer layer—and interface extends on two C layers. On the C face, the native (2x2) reconstruction saturates the DB states so that the first C layer can already exhibit graphene properties. The counterpart of this small interaction is that the long range orientation of the graphene layer is not imposed by the substrate. A balance has to be found between a long range order in the graphene layer that can be imposed by the substrate (strong interaction – Si face) and the preservation of graphene electronic structure (small interaction – C face). Decoupling the graphene layers from the substrate after the growth reveals an interesting point.

Epitaxial graphene on SiC have been patterned by microelectronics methods and desired band gap is created and tuned by laser irradiation (Singh *et al.* 2011).

Challenges in fabrication of epitaxial graphene by SiC method: Epitaxial growth yields very tiny amounts of graphene sheets and its surface cannot be modified, and hence it is not suitable for many applications like polymer nano-composites and will probably be limited to selected applications. Moreover, there is a need to eliminate the terraces or steps, i.e., the growth of the second or third layers at the edges of the terraces

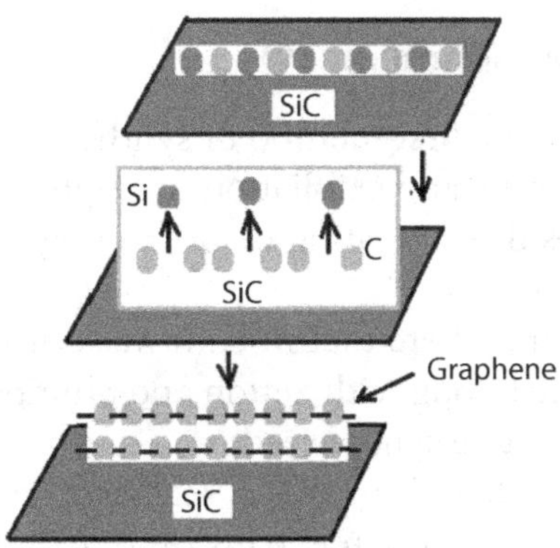

Figure 7.10 Showing that when SiC substrates are annealed at high temp., Si atoms selectively desorbs from the surface and the C atoms left behind naturally form FLG (few-layer graphene)

(which also strongly contribute to carrier scattering), an increase in the size of the crystallites and control of unintentional doping from the substrate and buffer layers.

7.5 Reduction of Graphene Oxide (GO)

Graphene oxide (GO) and graphene are different in structure. This difference is mainly due to the presence of (i) a large amount of functional groups attached to carbon planes and (ii) the structural defects within the plane. The existence of structural defects and functional groups decreases the electrical conductivity. That is why there is a need to reduce the GO. Reduced GO is being studied for many research as well as applications. When GO is reduced, it partly restores the structure and properties of graphene. Different reduction processes result in different properties of reduced GO or graphene, which in turn affect the final performance of materials or devices. By reducing GO, the elimination of functional groups as well as healing of structural defects can be achieved. The void that is created by removal of oxygen containing groups needs to be restored to a long-range conjugated structure, so that there are pathways for carrier transport within the graphene sheet. The healing of structural defects is being done by graphitization at high temperatures and epitaxial growth or CVD in the defective area with an extra carbon supply.

There are a number of ways to reduce GO to graphene. These methods are based on chemical, thermal or electrochemical treatment of GO. Usually exfoliated graphite oxide is used as precursor.

7.5.1 Thermal Reduction of GO

Reduction of GO was the first method of synthesis of monolayer flakes of graphene by rapid heating and exfoliation of graphite oxide (Boehm 1962). This method involves direct heating of GO to very high temperature in a furnace.

It is worth mentioning here that Boehm was one of the pioneers of the graphene research. He (along with Setton and Stumpp), characterized single graphene sheets that he synthesized.

Rapid heating (>2000°C/min) can exfoliate graphite oxide to graphene. This is due to the sudden expansion of CO or CO_2 gases into the spaces between stacked graphene layers. The oxygen attached to graphene as a functional group (attached to carbon plane) simultaneously gets decomposed into gases and creates huge pressure between graphene layers. McAllister *et al.* (2007) have calculated that 2.5 MPa is enough to separate two stacked GO platelets, whereas heating at 300°C generates 40 MPa. Thus the rapid heating exfoliates graphite oxide and reduces the functionalized graphene sheets by decomposing oxygen-containing groups. The graphene produced by this method are usually wrinkled and of small size. This is mainly because thermal reduction damages graphene sheet due to release of CO_2, and this also causes a loss in as much as 30% of GO and lattice defects (Schniepp *et al.* 2006). Significant impacts of the heating temperature of GO have been noted, e.g., at <500°C, the C/O ratio was no more than 7, whereas at 750°C, the C/O ratio was > 13 (Schniepp *et al.* 2006).

Challenges in fabrication of graphene by thermal reduction of GO: There are reports that heating damages the graphene platelets, because during heating pressure builds up between graphene layers and released carbon dioxide. Apart from damaging the structure of graphene, it also causes around 30% reduction in the mass of the GO, which creates imperfections and vacancies. It may also affect the mechanical strength of the graphene produced.

7.5.2 Hydrothermal Reduction of GO

Heating GO in distilled water at varying degrees for different lengths of time is called a hydrothermal dehydration route to convert GO to stable graphene solution (Zhou *et al.* 2009). In a hydrothermal process, overheated supercritical (SC) water acts as reducing agent and offers an alternative to the use of organic solvents as a reducing agent. In this process the aqueous solution of GO is autoclaved at 180°C for 6 hrs. After hydrothermal treatment the color of GO solution changes from yellow brown to black. This yields graphene precipitated as black powder.

During hydrothermal reduction of GO the water acts as a source of H^+. The advantage of this method is that apart from removing oxygen functional groups from GO, it repairs the aromatic structures as well. Moreover, by controlling the hydrothermal temperatures, physical properties of GO and tuneable optical limiting performance of graphene can be modified. In addition, its physiochemical properties can be widely changed with changes in pressure and temperature, which allows the catalysis of a variety of heterolytic (ionic) bond cleavage reactions in water.

pH of the aqueous solution of GO plays critical role in the type of graphene obtained. At high alkaline pH 11, stable graphene solution is obtained, whereas at acidic pH 3 the graphene obtained is highly aggregated, that cannot be dispersed even after adding concentrated ammonia solution. In this process, heating of H_2O promotes acid-catalyzed reaction of organic compounds because of a sufficiently high H^+ concentration compared to normal liquid phase.

Challenges in fabrication of graphene by hydro-thermal reduction of GO: The hydrothermal process produces steam at high temperature and pressure. At this condition, water yields H^+ ion, which helps in the reduction of graphene. In the hydrothermal process, no chemical is used except water, so the production of graphene becomes economical. However, this process can produce graphene of small sizes. Hence, this technique is suitable for small size graphene.

7.5.3 Solvothermal Reduction of GO

By solvothermal reduction methods, stable dispersion of graphene sheets are obtained, which is valuable for many applications. Different solvents have been tried for solvothermal reduction of GO. Some of them are discussed here:

i. ***N,N-dimethylformamide:*** This is a type of chemical reduction method. Basically it is performed by taking solvent (N,N-dimethylformamide) in a sealed container and heating it above its boiling point (153°C) at elevated pressure (Wang *et al.* 2009). Unlike hydrothermal reduction process where small amount of hydrazine was added as the reducing agent, here N,N-dimethylformamide is used. Solvothermal treatment is given at 180°C for 12 h. According to the Auger spectroscopy the C/O ratio of graphene is 14.3, which is much higher than that produced by hydrazine reduction at normal pressure.

ii. ***N-methyl-2-purrolidinone (NMP):*** Use of NMP does not require a sealed container (Dubin *et al.* 2010). Moreover the thermal energy used is lower than boiling point of NMP (202°C, 1 atm). Hence no extra pressure is present during reduction of GO. Deoxygenated GO is subsequently vacuum filtered. The C/O ratio of the Solvothermal reduced GO is only 5:15, which is much lower than that obtained by (14.3) N, N-dimethylformamide.

Challenges in fabrication of graphene by solvothermal reduction of GO: Solvothermal reduction methods can only achieve a moderate reduction of GO.

7.5.4 Chemical Reduction of GO

Chemical reduction of monolayer films of GO is used to synthesize graphene sheets by treating GO with hydrazine hydrate and maintaining the solution at 100°C for 24 hours. Chemical reduction is a very scalable method, but unfortunately the yield is relatively poor in terms of surface area and electronic conductivity

Reaction between GO and the chemical reagent usually realized at room temperature or moderate temperature results in the formation of graphene. Compared with the thermal method, chemical reduction is a cheaper method for the mass production of graphene. Different types of reducing agents have been tried over the years. Some of them are mentioned here:

i. ***Reduction of graphite oxide by hydrazine*** and its derivatives (hydrazine hydrate and dimethyl hydrazine) was first reported by Stankovich *et al.* (2006, 2007). These liquid-reducing agents are added to the water dispersed GO. This increases the hydrophobility and results in agglomerated sheets of graphene. It is dried to graphene powder. In another process GO is reduced while retaining the colloidal state in water by adding soluble polymers (Stankovich *et al.* 2006) as surfactant, or ammonia (Li *et al.* 2008).
ii. ***Reduction of graphite oxide by metal hydrides:*** Metal hydrides with strong reducing properties and very strong reactivity with water have been successfully used for the reduction of GO to synthesize graphene. These metal hydrides are sodium hydride, sodium boronhydride (NaBH4) and lithium aluminium hydride. It has been noted that $NaBH_4$ is more effective reductants of C = O species but has low efficiency in reducing epoxy and carboxylic group but alcohol group is not affected. It is more effective in reducing GO than hydrazine (Shin *et al.* 2009). It is suggested that sulphuric acid should be used after reduction by $NaBH_4$ at higher temperatures (180°C) to enhance the reduction effect (Gao 2009).

iii. ***Reduction of graphite oxide by ascorbic acid:*** Ascorbic acid, also known as vitamin C, has been attempted as a way to reduce GO by Fernandez-Merino *et al.* (2010). The graphene thus synthesized exhibits a C/O ratio of about 12.5 and a conductivity of 77 S/cm, which are comparable to those produced by hydrazine. The advantages of using ascorbic acid is that (i) it is not a toxic compound like hydrazine (ii) has higher chemical stability than $NaBH_4$ and (iii) in the colloidal state the reduced graphene does not agglomerate as has been observed when hydrazine is used.

iv. ***Reduction of graphite oxide by hydroiodic acid (HI):*** Hydroiodic acid (HI) is a highly reducing agent for the reduction of GO at low temperature to graphene films. GO was prepared from natural graphite flake by a modified Hummers method (Hummer and Hoffmann 1958) and assembled on a polyethylene terephthalate (PET) substrate. For reduction, GO is submerged into HI solution in a sealed cuvette placed in a thermostated oil bath. Thus produced graphene film is highly conductive (electrical conductivity as high as 298 S/cm and a C/O ratio above 12) and flexible. Pei *et al.* (2010) have suggested that conversion of GO to graphene film is based on the nucleophilic substitution reaction (Figure 7.11). The optimal temperature for reduction of GO films is 100°C using 55% HI when reduced for 1 h. Both

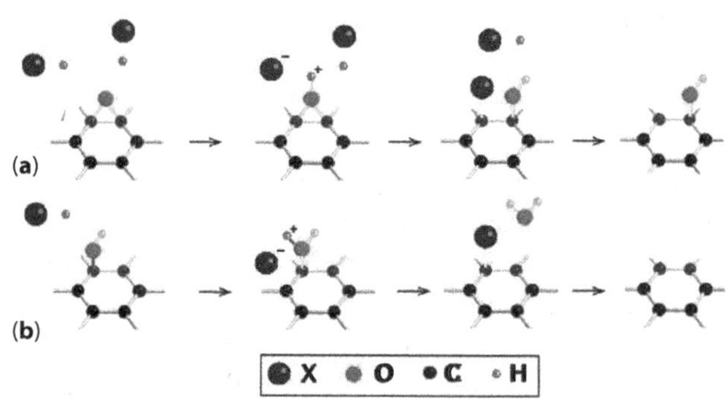

Figure 7.11 Possible reaction mechanism of GO reduction by hydroiodic acid (a) Ring-opening reaction of an epoxy group (b) Halogenation substitution reaction of a hydroxyl group. The substituted halogen atoms are expected to be easily eliminated from the carbon lattice. (X = iodine or bromine). Copyright © 2010 Elsevier Ltd. All rights reserved.; Courtesy Pei *et al.*, *Carbon*, 48, Issue 15, 2010, Pages 4466–4474.

electrical conductivity and a C/O ratio of this graphene film is much higher than films reduced by other chemical methods. Moreover, reduction by this method not only maintains good integrity and flexibility but also improves the strength and ductility. It is a valuable method for the large-scale production of graphene.

7.5.5 Electrochemical Reduction of GO

Very high quality graphene almost similar to pristine graphene is produced by the electrochemical reduction of GO. The basic process of electrochemical reduction of GO involves coating a thin layer of GO on a metal substrate such as indium tin oxide, plastic or glass, to which electrodes are attached at each end. An inert electrode is placed opposite the film in an electrochemical cell and reduction occurs during charging of the cell, thus creating a circuit through the GO. Then by linear sweep, voltammetry or cyclic voltammetry of the GO in an aqueous sodium phosphate buffer is carried out in the range of 0–0.1 V (respect to a saturated calomel electrode) to a GO-modified electrode in a 0.1 M KNO_3 solution. At 0.6 volts, reduction begins, and maximum reduction is observed at 0.87 volts (Ramesha and Sampath 2009). The whole process takes place at room temperature. During electrochemical reduction of GO, oxygen functionalities are removed by the electron exchange between GO and electrodes. The advantage of this method is that since no chemical agent or dangerous reductants, e.g., hydrazine are used, no toxic by-product is generated. Zhou *et al.* [110] have reported that the pH value of the buffer solution controls the potential needed to realize the reduction and low pH is suitable for the reduction of GO. This suggests that the H+ ions participate in the reaction.

Other primary benefits of this technique are that there are no hazardous chemicals used, meaning no toxic waste to dispose of.

Challenges in fabrication of graphene by electrochemical reduction of GO: One of the challenges is the low scalability of this technique due to the difficulty in depositing graphene oxide onto the electrodes in bulk form. Moreover, as compared to scotch-tape method, the quality is low because the oxidation protocol introduces permanent defects due to over-oxidation. Electrochemically reduced GO has very high carbon to oxygen ratio.

7.5.6 Reduction of GO by Hydrogen Plasma

In this method reduction of GO is done by exposing GO hydrogen plasma for a few seconds (Figure 7.12). It is done at 150°C and with atmospheric

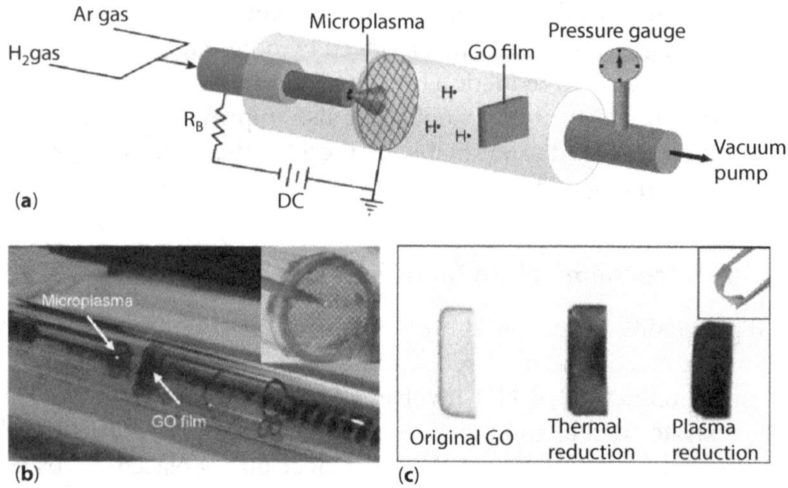

Figure 7.12 (a) Schematic diagram of experimental setup used for remote, atmospheric-pressure plasma-assisted reduction of GO thin films. DC microplasma housed inside a quartz tube is formed between a negatively biased stainless-steel capillary tube and a grounded stainless-steel mesh electrode. Gas mixtures of Ar and H_2 are introduced into the microplasma to generate atomic hydrogen (H•). Thin films of GO were drop cast on substrates and placed 1.5 cm from the microplasma to avoid ion and electron bombardment. (b) Photograph of remote microplasma, operated in Ar/H_2 (50/50) at a discharge current of 10 mA and GO film (inset: close-up of microplasma). (c) Photographs of GO films deposited on PET substrates before reduction (left panel) and after thermal treatment at 150°C (middle panel) or plasma treatment at 150°C (right panel). All treatments were performed in an Ar/H_2 atmosphere (50/50) at 1 atm (inset: flexibility of GO films after plasma-assisted reduction is demonstrated). Copyright © 2012, American Chemical Society Copyright © 2012, American Chemical Society, Courtesy: Seung Whan Lee, *et al. J. Phys. Chem. Lett.*, 2012, 3 (6), pp 772–777, doi: 10.1021/jz300080p.

pressure. In this method hydrogen gas along with Ar in the ratio of 50:50, is continuously dissociated in a microplasma at a discharge current of 10 mA to generate atomic hydrogen and flows towards thin films of GO deposited on a substrate, which is placed 1.5 cm from the microplasma to avoid ion and electron bombardment.

7.5.7 Reduction of GO by Xenon Flashtubes

Reduction of GO by exposing GO to strong pulse light, such as those produced by xenon flashtubes, is completed. It is a chemical-free flash reduction process; where flash-light from a photographic camera is used to instantaneously trigger the deoxygenation reaction of GO by photothermal heating. Flash reduction can be carried out in air or under N_2 atmosphere inside a glove box. This method has been used to fuse graphene and a polymer, and also to create a composite. Cote *et al.* (2009) have suggested that using a photomask, conducting patterns such as interdigitated electrode arrays can be readily made on flexible substrates.

For xenon flash bulb reduction of GO, thin films were prepared by filtering stock solution of GO through a 47 mm diameter anodized aluminum oxide (AAO) membrane having pore size of 0.02 μm. Filtered GO is dried in air to remove any residual water molecule and then the thin film of 1μm thickness is peeled off the AAO.

The typical flash energies applied to the samples through flash tubes is around 0.1–2 J/cm^2. In order to improve the uniformity multiple flashes of lower energy is applied until the desired color was reached.

7.5.8 Reduction of GO by an Expansion-Reduction Agent

Combining GO with an expansion-reduction agent such as urea and then heating the solution to cause the urea to decompose and release reducing gases. It is followed by heating (600°C) in an inert gas (nitrogen) environment. It is subsequently cooled and graphene is obtained as a solid by-product. (Wakeland *et al.* 2010). A combination of urea as a fuel and ammonium nitrate as oxidizer (Satya Kisore 2008), along with GO was heated at lower temperature (250°C) has also been used for the same purpose, i.e., for the formation of thin-graphene sheets.

7.5.9 Photocatalytic Reduction of GO

The feasibility of reduction of GO by photocatalytic properties of different semiconducting particles has been attempted. It includes TiO_2 (Williams *et al.* 2008), ZnO (Williams *et al.* 2009) and $BiVO_4$ (Ng *et al.* 2010)

a. TiO2
   This is the most commonly used nano particle for photocatalytic reduction of GO. The photocatalytic properties of semiconducting TiO_2 particles are well known and thoroughly reviewed (Kamat 1993). The procedure involves dispersion of GO in ethanol forming a colloidal state. The process involves UV irradiation of colloidal TiO_2 dispersed in ethanol containing suspension of GO. When TiO_2 is UV

irradiated, charge separation occurs on the surface of TiO_2 particles. This mixture is originally light brown in color but as the reduction of GO takes place the color changes to dark brown and then black. In the presence of ethanol the holes from the TiO_2 are scavenged to produce ethoxy radicals, thus leaving the electrons to accumulate on the surface of TiO_2 particles. The accumulated electrons interact with GO sheets to reduce functional groups (Williams *et al.* 2008).

Following the same idea, some other materials with photocatalytic activity such as ZnO and $BiVO_4$ have also been reported to achieve the reduction of GO.

b. ZnO

ZnO nanoparticles have also been used to photocatalytically reduce the GO. The process is very similar to that takes place with $TiO_{2.}$ Photo-excited ZnO nanoparticles interacts with GO sheets suspended in ethanol. GO sheets then undergo photocatalytic reduction. The luminescence quenching of the green emission of ZnO serves as a probe to monitor the electron transfer from excited ZnO to graphene oxide. ZnO nanoparticles anchored on GO can be used to design semiconductor-carbon nanocomposites for catalytic applications (William *et al.* 2009).

c. Bismuth Vanadate ($BiVO_4$)

$BiVO_4$ has also been used to reduce GO by photocatalytic action in presence of visible light. Ng *et al.* (2010) have demonstrated that "$BiVO_4$-graphene generates higher photocurrent density in visible light than that generated by the TiO_2 system in UV light. Moreover, significant H_2 and O_2 evolution is seen in the $BiVO_4$-RGO photoelectrochemical water splitting cell, whereas negligible gas evolution is observed in the pure $BiVO_4$ cell, showing the great advancement of the synthesized $BiVO_4$-RGO composite for water splitting".

7.5.10 Multi-Step Reduction

It has been observed (Eda *et al.* 2010 and Gao *et al.* 2009) that the single process induced reduction strategies like chemical reduction by hydrazine may not yield total reduction of GO. Hence, there is a need to further improve the reduction process. For this purpose multi-step reduction has been proposed.

i. Eda *et al.* (2010) found that a combination of hydrazine reduction and low temperature thermal annealing at 200°C produces graphene sheet with better conductivity than that produced by only thermal annealing at 550°C. Higher conductivity is needed for flexible devices attached to polymer substrates.
ii. Use of chemical reagents for reducing GO is often selective and cannot eliminate all oxygen-containing functional groups. Therefore, Gao *et al.* proposed a three-step reduction process:

   Step 1: Deoxygenation with $NaBH_4$, this will eliminate any ketone, lactol, ester and most alcohol groups.

   Step 2: Dehydration with concentrated sulphuric acid is useful because it forms alkene bonds that are part of a conjugated sp^2 carbon network.

   Step 3: Thermal annealing in Ar/H_2 atmosphere at 1100°C for 15 min to reduce the oxygen content to less than 0.5 wt.%, which is close to the value in graphite powder. This treatment has yielded graphene having a C/O ratio higher than 246, which is the highest value reported until now, but the electrical conductivity of this film is only 202 S/cm, much lower than that achieved by direct annealing in an Ar atmosphere at the same temperature reported by Wang *et al.* 2009.

Challenges in Fabrication of graphene by Multi-step Reduction of GO: Many strategies to reduce GO has been attempted but scientists are still not sure whether, all the functional groups of a GO sheet are fully eliminated or the lattice defects formed during oxidation can be restored during reduction. Another question that troubles is whether the reduction process decreases or increases the defect density in a graphene sheet? However, some of these techniques have produced very high quality graphene that are similar to pristine graphene, but these multi step approaches can be complex or time consuming to carry out.

7.6 Arc-Discharge Method

In quest for large scale production of high quality graphene for successful application, the arc-discharge method has also been tried (Choucaire *et al.* 2009; Li *et al.* 2010; Wang *et al.* 2010). This method has certain unique advantages, such as it does not need any metal catalysts, process is facile and time-saving, and grams of high quality single, bi and multilayer graphene of a few 100 nm to little μm size are produced. This method is a top-down approach of synthesizing graphene from graphite. During synthesis of SWCNT by arc-discharge method, some fragments of carbon dots or graphene dots are also formed.

The arc-discharge setup comprises of an electric arc oven where anode and cathodes of pure graphite rods are placed through which DC current (100–150 A) is passed. As the rods are brought close together, discharge occurs resulting in the formation of plasma and soot is generated. The soot deposited on the wall of chamber is graphene. By this method, boron-doped graphene can be synthesized by carrying out discharge in the presence of a mixture of H_2 and B_2H_6 (Subrahmanyam *et al.*, 2009). N-doped graphene is synthesized by carrying out the discharge in the mixing atmosphere of He and NH_3 (Li *et al.* 2010a). However, pure graphene is synthesized in presence of H_2. Graphene is also synthesized in the presence of air (Wang *et al.* 2010). Wu *et al.* have developed an arc-discharge method using a buffer gas containing carbon dioxide to synthesize few-layered graphene.

Challenges in Arc Discharge Method: Are due to low yield of graphene with some carbonaceous impurities.

7.7 Solvothermal Method

It is a simple and possible mass production process, in which sodium and ethanol with molar ratio of 1:1 are taken in a Teflon-lined container and heated at 220°C for 72 h to form a graphene precursor.

At this temperature the alcoholic solution becomes increasingly saturated with the metal alkoxide, and thus auto-generated pressure (~1×10^{-2} bar) is created. The free alcohol gets encapsulated into the metal alkoxide in a clathrate-like structure.

A clathrate consists of a lattice that traps or contains molecules. The word *clathrate* is derived from the Latin *clatratus* meaning *with bars or a lattice*. According to IUPAC, "Clathrates are inclusion compounds in which the guest molecule is in a cage formed by the host molecule or by a lattice of host molecules."

This graphene precursor is rapidly pyrolyzed. Nucleation of sheets occurs around the regions with rich in ethanol. The product is then washed, but it is not purely the ethoxide-ethanol clathrate; some metal hydroxide may also form in the presence of water. Finally the product is vacuum filtered and dried in a vacuum oven at 100°C for 24 h.

7.8 Substrate-Free Gas Phase Synthesis Of Graphene

This is a rather new method developed by Dato *et al.* (2008). By this method, clean, highly ordered, free-standing graphene sheets can be produced continuously at ambient conditions. This method uses an aerosol of ethanol droplets and argon gas, which is directly sent into atmospheric-pressure microwave-generated argon plasma. Ethanol droplets within 10^{-1}s, get evaporated and dissociated in the plasma, forming solid matter (Figure 7.13). This method does not need post-synthesis treatments, such as chemical reduction, dispersion in liquids, or thermal annealing, etc. The graphene can be directly mounted onto a TEM grid or deposited on Si substrate. Dato *et al.* have claimed that by this method, within a few minutes, a milligram amount of graphene can be produced and it can be easily scaled up to commercial production level.

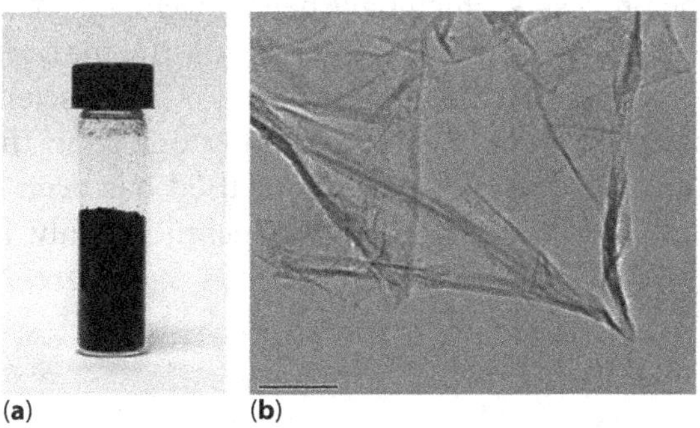

Figure 7.13 (a) A 6 cm tall vial containing a 100 mg sample. In experimental setup, samples were obtained at a rate of 2 mg/min.(b) A typical low-magnification TEM image of synthesized graphene sheets. Scale bar is 100 nm. Copyright@rsc.org Royal Society of Chemistry; Courtesy: Dato *et al. Chem. Commun.*, 2009, 6095–6097 doi: 10.1039/b911395a.

7.9 Other Growth Methods

Nowadays many publications appear each month claiming new growth methods for synthesis of graphene. However, very few of them are likely to be used for viable commercial production in the near future, e.g. (i) Molecular beam Epitaxy to grow chemically pure graphene, but its large scale production may not be possible because of its much higher production cost than CVD methods (ii) another method is surface-assisted coupling of molecular monomer precursors into linear polyphenylenes with subsequent cyclo-dehydrogenation way to create high-quality graphene nanoribbons. (iii) An expensive potential technique by laser ablation is also being worked out, which may compete with spray coating of graphene. But this does not mean that it should not be considered; rather, it should be researched further. There is a need to look for more chemistry-driven bottom-up approaches to synthesize graphene.

7.10 Summary

This chapter has been devoted to discuss the various methods used for the production of graphene at bench scale and attempts made to produce it on the large scale. Care has been taken to discuss the various methods used for synthesis especially because the purity of material's requirements may vary from lab to lab depending upon the specific requirements. In addition, every research lab has only specific facilities and hence each lab cannot adopt all the processes which have been developed so far. Considering these two aspects, efforts are made to highlight the maximum number of techniques used for synthesis of graphene from GO, so scientists can select the most suitable technique that could be performed in their lab. This is important because to date, not a single method has been claimed as the best, so that every lab could work on that technique only. It is hoped that this approach would be acceptable by various readers/researchers in this field.

8

Direct Transfer or Roll-To-Roll Transfer of Graphene Sheet onto Desired Substrate

The life-blood of our business is that R and D spent. There is nothing that flows through a pipe or down a wire or anything else. We have to continuously create new innovations that let people do something they didn't think they could do the day before.

Steve Balmer

8.1 Introduction

Successful applications of graphene for commercial purposes depends on suitable methods of synthesis of good quality graphene, the transfer of graphene on deterministic substrate, and the doping of graphene as required for the particular application. Though the ideal way would have been to synthesize it on a needed substrate, so far we have not reached this goal; we are depending on the transfer of graphene synthesized on Cu foil, Ni or SiO_2/Si to other substrates either by using poly (methyl methacrylate) (PMMA) as a sacrificial layer or by dry-methods based on a polydimethylsiloxane (PDMS) an elastomeric stamp. This crucial method is necessary for electrical current to flow through graphene devices because conducting substrate/catalyst hinders it. Therefore, there is a need to transfer graphene to an insulating surface (Kedzierski *et al.* 2009).

Some efforts at lab-scale are made to avoid using metal foil, so that the process of making large size graphene on a commercial level becomes economical. Hence, there is a need to develop a process which does not use metal foil. Moreover, the transfer process developed has to be such that it can be scaled up to yield graphene films in a uniform, effective manner and industry-compatible method. The roll-to-roll process may be an optimal solution, opening up new possibilities for future applications.

Currently, graphene of various sizes and types are being synthesized, e.g. a few mm of single crystals of graphene (Zhou *et al.* 2013) to meters of poly-crystal sheet of graphene (Bae *et al.* 2010). Transferring different types of graphene from metal to target substrate without damaging it is a critical step. People have attempted several methods for transfer. In this section we discuss various transfer processes of synthesized graphene sheet onto desired substrates, and the potential for scaling up so that it can be used in mass-scale production. Whichever transfer method is used, basically all involve following steps:

i. Coating a thin protective film with an adhesive on the graphene grown on Cu foil, Ni, Si/SiO_2 substrate or any other substrate.
ii. Etching away or removing the substrate used using etchant, e.g., Cu foil is removed by treating it with aqueous 0.1 M ammonium per sulphate solution $(NH4)_2S_2O_8$ (Jo *et al.* 2005) or $FeCl_3$ solution. Thermal or electrochemical treatment is completed to remove the polymer.
iii. Finally, transferring the graphene from the polymer support onto the desired substrate directly, or by using a roll-to-roll system.

It must be noted here that graphene is synthesized by many different methods onto different substrates (See chapter 7). Transfer methods of graphene are bound to vary for differently synthesized graphene, e.g., graphene synthesized by CVD method can be processed by roll-to-roll transfer method, but it may not be suitable for graphene synthesized by other techniques such as mechanical exfoliation or chemical exfoliation dealing with multi-layer graphite as precursor material. Similarly, not all transfer techniques are applicable for CVD graphene.

Monolayer graphene poses more problems during transfer because the delicate monolayer can get damaged during transfer or the alignment between target substrate and graphene may not be perfect; moreover, if the transfer procedure is performed in aqueous solution, there are chances of liquid getting trapped within the interfacial space between graphene and the target substrate.

8.2 Direct Transfer of Graphene by Etching and Scooping Method

This is a simple two-step method. The first step is to get rid of catalyst film (substrate), on to which the graphene is synthesized, by etching. For this, the sample is submerged in an etchant. The etchant is selected as per the substrate/catalyst film that is to be removed. After it is dissolved, the graphene floats to the surface of the liquid. Graphene, being hydrophobic, does not lose its form.

The second step is to scoop away the floating graphene on the desired target substrate. Scooping has to be done very carefully so that a uniform sheet of graphene is taken up on the surface. Graphene sheets are very fragile and require careful handling. Once this step is completed, the scooped

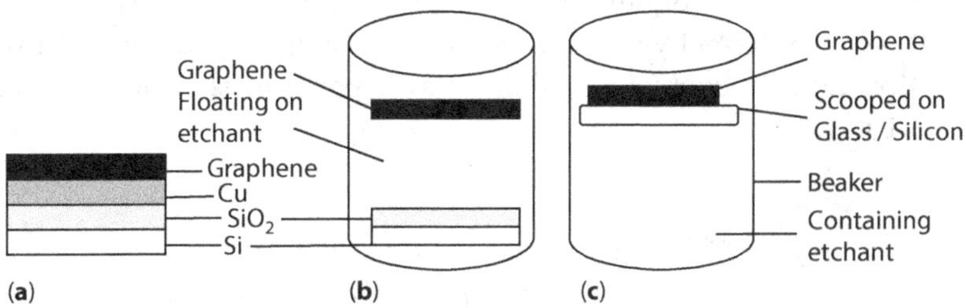

Figure 8.1 Schematic diagram of the etching and scooping method (a) The CVD-grown graphene on Cu catalyst, (b) The graphene sample is floating when dipped into the metal etchant, (c) Floated graphene is scooped on a glass/Si.

away (on some substrate like glass) graphene is dried. A schematic representation of this method is given in Figure 8.1.

The challenges faced in this method: During scooping (as mentioned above), it has to be done very carefully or otherwise sheets may break. The other problem is that if etchant gets in between the graphene and the target substrate during scooping, it behaves like a dopant, which may enhance the carrier scattering, thus degrading the electrical properties.

8.3 Direct Transfer of Graphene by Etching and Scooping Method Using a Graphene Protecting Media

This method is similar to that of the etching and scooping method. However, in this method, prior to etching, a protective polymer is placed on the graphene sheet. The most common media that is used for protecting graphene is PMMA, PDMS, PC (Poly(bisphenol A Carbonate) and also using thermal release tape.

8.3.1 PMMA

PMMA or Poly(methyl methacrylate) is a transparent, thermoplastic, lightweight, strong, shatter-resistant material with a 1.17–1.20 g/cm^3 density. It has good impact strength, and it ignites at 460°C. When burned, it forms carbon dioxide, water, carbon monoxide and formaldehyde. A 3 mm thick PMMA transmits up to 92% of visible light and reflects about 4% light from both the surfaces because it has a refractive index of 1.4914 at 587.6 nm. It can filter UV light at a < 300 nm wavelength. PMMA can be dissolved in many organic solvents. It has a maximum water absorption ratio of 0.3–0.4% by weight. Its tensile strength decreases with water absorption and its coefficient of thermal expansion is relatively high at (5–10)×10^{-5} K^{-1}.

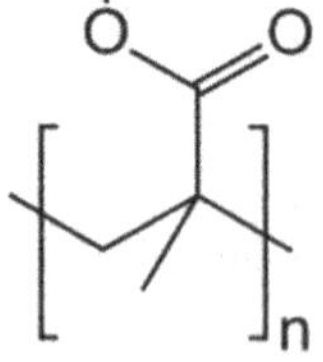

PMAA (Poly (methyl methacrylate)

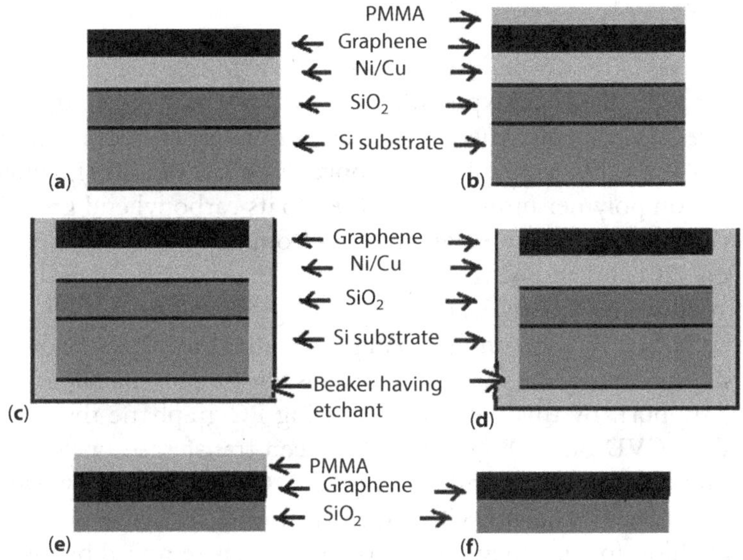

Figure 8.2 Schematic diagram of graphene transfer with PMAA as a support (a) The CVD grown graphene on Ni or Cu catalyst (b) A PMAA layer is spin-coated on top of graphene, (c) The graphene is submerged into the metal (Ni or Cu) etchant (d) The Ni or Cu is etched and the graphene is floating with PMAA on the etchant surface, while the remaining SiO_2 and Si substrate sinks to the bottom of the beaker. (e) The floating graphene/PMAA is transferred onto SiO_2 substrate. (f) The PMAA top layer is removed by acetone or other PMAA remover and graphene remains on SiO_2.

PET using PMMA as the support polymer shows the lowest sheet resistance values (125 Ω/ϒ), whereas the sheet resistance for the same graphene transferred onto a SiO_2 substrate also using PMMA is 350 Ω/ϒ. PMAA is usually applied on the graphene/metal surface by spin-coating. After the deposition of graphene on the target substrate, PMMA is removed either via high-temperature annealing (350–500°C) or by dissolving it in acetone or in an etchant.

PMAA is supposed to be a better support for graphene in comparison to thermal release tape or PDMS; when PMAA is used, the number of cracks and defects are lower. That is why when graphene films are transferred with PMMA, they show the lowest sheet resistance. Moreover, the electrical properties of the transferred films also depend on the substrate on which they are deposited.

8.3.2 PC (Poly (bisphenol A Carbonate)

Because PMMA is an additive polymer that is difficult to decompose thermally, it swells when exposed to a solvent and difficult to disentangle with aggressive solvents. Therefore, as an alternative to PMMA, scientists have tried using PC as graphene support (Wood *et al.* 2012), which is a condensation polymer more reactive (due to its carbonyl end group), and therefore easier to decompose and remove from the graphene more cleanly (Park *et al.* 2012) than PMMA.

The method involves CVD synthesis of graphene on Cu. PC is used after a modified RCA cleaning (Liang *et al.* 2011) for transfer of layered graphene structures to arbitrary substrates with a poly(bisphenol A carbonate) support. By this method (i.e., using PC graphene that was synthesized by CVD on Cu), graphene has been transferred onto 90 nm of $Si0_2/Si$ and also on cleaved mica (VI grade). Cu was etched by overnight prolonged treatment with $FeCl_3$. The PC is floated off by keeping the graphene in chloroform. When PMAA is applied it is removed by dissolving in an etchant.

8.3.3 Transfer on Pre-Stretched Substrate, PDMS

If PDMS and thermal tapes are used, they are not dissolved; they are peeled off. Since PDMS is to be peeled off, it is important that the interface bonding between graphene and substrate is stronger than between PDMS and graphene. This is necessary so that when PDMS is peeled off, the graphene remains attached to the substrate. That is why the PDMS stamp is attached to graphene when it is still sticky or partially cured. Before being peeled off, PDMS is completely cured, which reduces the adhesion strength of PDMS and supports easy peeling. To increase the bonding between graphene and surface of the target substrate, often substrate is activated by plasma. (Miao *et al.* 2011).

The challenges faced in this method: Peeling of the PDMS without creating any damage to the graphene film. This becomes more serious if the graphene film is of a larger size. Therefore, care must be taken when peeling off the PDMS.

8.3.4 Direct Transfer of Graphene onto Flexible Polyethylene Terephthalate (PET)

In this method, graphene synthesized on Cu foil by CVD method is directly transferred onto PET. PET is a transparent polymer. The transfer process is carried out using a lamination machine. After transfer, the Cu foil is dissolved by 1 M $FeCl_3$ solution in 3% HCl; thus, a ready-to-use graphene on PET substrate is produced by a single-step approach. As large as 40" in diameter and >98% graphene has been synthesized by this method (Vlassiouk *et al.* 2013). The characteristics of this graphene are very encouraging for its various applications requiring transparent conducting electrode materials. The sheet resistance was 1-3kX/sq., which is similar to smaller samples reported by Li *et al.* (2009), and the resistance can be further reduced by doping (Shi *et al.* 2010).

A slightly modified method using a lamination process was developed by Martin *et al.* (2013). Using this method, CVD-deposited graphene was transferred onto different flexible substrates such as polytetrafluoroethylene filter membranes, PVC, cellulose nitrate/cellulose acetate filter membranes, polycarbonate, paraffin, polyethylene terephthalate, paper or cloth (Martin *et al.* 2013). In this method, PMMA is not used as an intermediate transfer membrane. But because successful transfer depends on the substrate's hydrophobicity and good contact between the substrate and graphene, hydrophobic substrates with low glass transition temperature are suitable for transfer. For substrates like paper or cloth that do not meet these criteria, a polymer such as PMMA can be used as a surface modifier or as an adhesive to ensure successful transfer.

The procedure involves CVD deposition of graphene on both the sides of Cu foil to make a graphene/Copper/graphene stack. A protective paper is placed on the stack. This stack is cut and then placed between target substrate PET and pressed to provide initial graphene-substrate contact. A thorough cleaning of the target substrate is a must. It is done by cleaning the substrates with iso-propanol and blow-drying it with a nitrogen gun. However, if substrates like paper and cloth are used, it is only dusted with a nitrogen gun. With this setup, first a protective sheet of weighing paper is placed on top of the target substrate and then between two PET films, a whole stack of PET/substrate/(G/Cu/G)/paper/PET sandwich is laminated. PET films are used to prevent direct contact between Cu foil and the rollers of lamination machine and protective paper is used to prevent the copper foil from adhering to the PET films, because during lamination the temperature goes above 100°C and PET becomes viscoelastic. Lamination is done by inserting the graphene stack into a hot/cold lamination machine. Due to lamination, enough pressure is provided to create close contact between the graphene/copper interface and the substrate. The laminating material, soft Paraffin (Parafilm), has a low melting point; it melts during heating and this helps in keeping the graphene

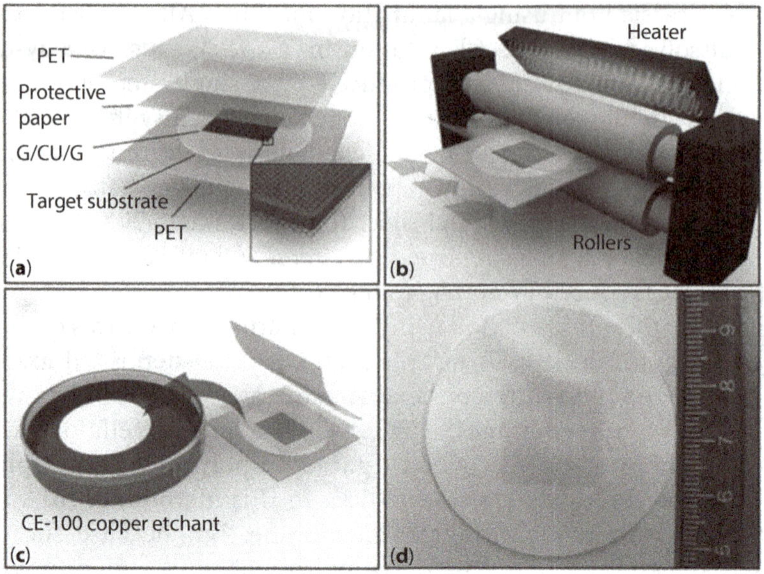

Figure 8.3 Schematic diagram of the direct transfer method of graphene via lamination. (a) Copper foil with CVD graphene grown on both sides (G/Cu/G) is placed in between the target substrate and the protective paper. This stack is then put between two PET films. (b) The PET/substrate/(G/Cu/G)/paper/PET sandwich is inserted into the hot/cold lamination machine. (c) The PET films and the protective paper are then removed and the remaining substrate/graphene/copper stack is floated on a copper etchant solution for 15 min. (d) The graphene/substrate is rinsed in DI water and blowdried with nitrogen. In this picture, graphene is on a Teflon filter. The ruler is scaled in centimeters. Courtesy: PNAS; Martins, L. G. P., Y. Song, T. Zeng, M. S. Dresselhaus, J. Kong, and P. T. Araujo. "Direct Transfer of Graphene onto Flexible Substrates." *Proceedings of the National Academy of Sciences*,110, no. 44 (October 14, 2013): 17762–17767. As Published http://dx.doi.org/10.1073/pnas.1306508110

sheets attached to the substrate during the etching process. Prior to etching, PET films and the protective paper are removed. The remaining substrate/graphene/copper stack is floated on a copper etchant ($FeCl_3$) solution for 15 min. The graphene on the substrate is then rinsed with deionized water and blowdried with a nitrogen gun. Removed PET can be reused. By repeating the whole process, multiple layers of graphene can be obtained.

The challenges faced in this method: Since this method involves deposition of graphene on both the sides of Cu by CVD process, care is needed to maintain uniform thickness of graphene film as well as the number of layers deposited on each side of the Cu substrate. If this can be achieved, the technique will be suitable for commercialization.

8.4 Roll-to-Roll Synthesis and Transfer of Graphene

Roll-to-roll transfer is a standard method used in various industries such as food and electronics, where fast, continuous and high productivity is required. The roll-to-roll method of transfer of graphene produced by CVD has shown outstanding scalability/processibility at a commercial level. Moreover, it is observed that roll-to-roll processing is inexpensive. The roll-to-roll method involves both (i) synthesis on a desired substrate as well as (ii) transfer to a target substrate both by using Rolls. In the roll-to-roll CVD process, graphene is continuously synthesized on the metal foil as it passes through a high temperature reactor using several guide rollers.

8.4.1 Roll-to-Roll Continuous Transfer Using Thermal Tape

This method was reported by Bae *et al.* (2010). It is called roll-to-roll because first the graphene film is grown on roll-type flexible Cu substrates by CVD at 1000°C under a controlled flow of hydrogen and pressure. A roll-type substrate not only supports its fitting into tubular furnace, but also maximizes the homogeneous deposition of graphene film on the flexible Cu roll that has been annealed. The roll that Bae *et al.* used was diagonally 30″ in scale.

Prior to transfer of the graphene layer, a thermal-release tape is placed between two rollers by applying soft pressure (~0.2 MPa), and then Cu layer is etched or removed by electrochemical reaction with a particular Cu etchant filled in a plastic bath. Traces of etchant left are removed by distilled water wash. After that, the thermal tapes are detached from graphene by thermal treatment, i.e., a mild heat treatment (90~120°C) for 3~5 min. Then the graphene is transferred to the target substrates (Figure 8.4). This graphene sheet is then transferred layer by layer onto flexible substrates using the roll-to-roll method. To prepare multilayer graphene sheets, this process should be repeated on the same substrate. Care is needed to clean the copper sheet such that the deposition of graphene takes place without any defects. Thermal release tape should be able to hold the graphene sheet while copper is chemically removed.

The challenges faced in this method: Graphene transfer by thermal release tape results in low-residue graphene; moreover, it is highly unlikely to result in continuous coverage.

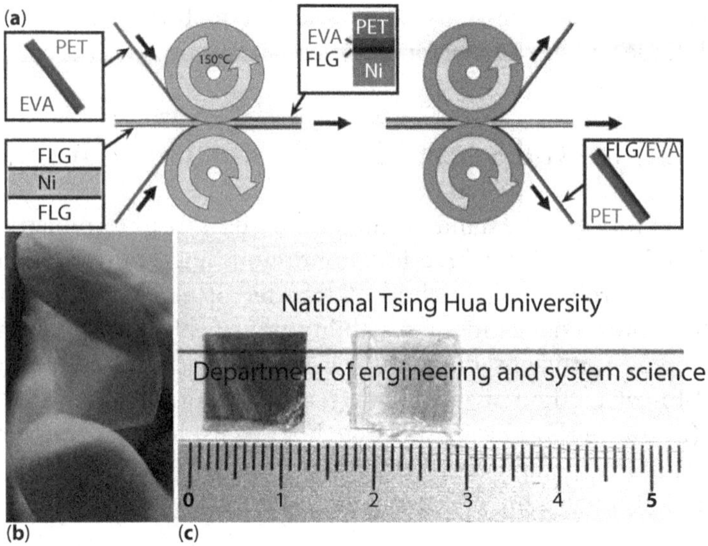

Figure 8.4 (a) A diagram to show the roll-to-roll process for the transfer of FLG from Ni foil to EVA/PET substrates. (b) A 2 × 3 cm^2 transparent flexible FLG/EVA/PET sample (c) Two 1 × 1 cm2 transferred samples with different transmittances. .Courtesy: Copyright © 2010 Elsevier Ltd. All rights reserved Zhen-Yu Juang *et al.*, graphene synthesis by chemical vapor deposition and transfer by a roll-to-roll process; *Carbon* 48, 3169(2010), doi:10.1016/j.carbon.2010.05.001.

8.4.2 Roll-to-Roll Transfer on to Ethylene-Vinyl Acetate Copolymer (EVA) Coated Transparent Poly-Ethylene Terephthalate (PET) Sheets by Hot Press Method

The principle of this method is the same as that of the transfer of graphene onto flexible polyethylene terephthalate (PET), mentioned earlier. However, in this method, two polymer as substrate and a roll-to-roll process was used. This method is appropriate for transfer of single (SLG) and a few layers of graphene (FLG). SLG to FLG of centimeter scale is grown on rolls of flexible transition metals Ni foil (30μm thick) by CVD method. Flexible substrate for graphene synthesis is necessary for the use of a roll-to-roll transfer process. Prior to growing graphene on the Ni roll, it was inserted in a tubular furnace and annealed at 900°C for 10 min under a H_2/Ar (10 and 400 sccm, respectively) atmosphere. Then, a mixture of gases (CH_4/H_2/Ar at the rate of 10, 100 and 400 sccm, respectively) was allowed to flow in the furnace to synthesize graphene kept at 750 Torr pressure. Maintaining the same pressure and flow of gas mixture, the furnace is quickly cooled to room temperature.

For the roll-to-roll transfer, graphene layers on Ni foil (FLG/Ni) and target substrate (EVA/PET) were pressed together with hot rollers (150°C). The graphene layer is now sandwiched between EVA/PET and Ni sheet. The EVA layer acts as a viscose between the PET and FLG. To separate the PET/EVA/FLG layers from the Ni surface in a uniform manner, the hot-rolled PET/EVA/FLG/Ni sheet is then passed onto a cold roller (room temperature) using a constant rolling speed. This method of roll-to-roll transfer process was developed by Juang *et al.* of National Tsing Hua University, Taiwan. However, it must be mentioned here that by this method one gets graphene sheets of undesirable high sheet resistance and low transmittance. Hence, there is a need to research the doping of these samples.

Challenges in this method: The design of the CVD unit, such that while the copper sheet moves by roll-to-roll system, there is no leakage of hydrogen and methane gas. In addition, care will be needed so that oxygen does not enter the system; otherwise there can be an explosion at high temperatures with hydrogen.

8.4.3 Roll-to-Roll Transfer Using Photo-Curable Epoxy Resin onto a PET Film

Because the roll-to-roll transfer method depends on the graphene synthesized by CVD method on rolls of substrate using high temperature, it often faces problems like synthesis of graphene without defects and multilayer graphene. Yamada *et al.* (2012) have synthesized 297 mm wide graphene on Cu foil on a large scale at a relatively low temperature of 400°C by assisting the dissociation of methane using microwave plasma, but sheet resistance of such graphene film was higher than 10 kΩ/ϒ. To combat these problems and produce high quality graphene, Kobayashi *et al.* (2013) designed a roll-to-roll continuous graphene synthesizing CVD set up and used selective Joule heating to heat a copper foil suspended between two current-feeding electrode rollers at ~1000°C.

A reverse gravure coating of a photo-curable epoxy resin onto a PET film and bonding to the graphene/copper foil is completed. The epoxy resin is then cured and Cu foil is removed by spraying an etching agent like $CuCl_2$ solution. This finally gives a structure "graphene/ epoxy/PET sheet." The use of epoxy resin significantly improves the transfer process as compared to the conventional two-step transfer process, i.e., first transferring the graphene film to a support film and then moving it to the target substrate. The set up is unique because in the vacuum steel chamber, a pair of winders and a pair of current-feeding electrode rollers are used, and the copper foil is suspended between the two rollers and the heater. The system is such that all rollers are electrically isolated from each other before loading the copper foil (230mm wide, >100m long, 36 μm thick, >99.9% pure) and all of the current flows through the suspended copper foil. During synthesis, the pressure is maintained at 1000 Pa under a continuous flow of CH_4 (450 sccm) and H_2 (50 sccm). This low-pressure CVD synthesis suppresses the sublimation of Cu at high temperature. Another important consideration is the winding of Cu foil on the roll-to-roll system that is done for 100 min at a rate of 0.1m/min. This is helpful in keeping the CVD process stable throughout the process. This process takes more than 16 hr. Initially, approximately 50% of the graphene sheet is grown at Tg ≈ 950°C and then at J¼83 A/mm^2. The temperature is estimated from the visible spectrum of emission from the copper foil.

8.5 Apparatus Used for Roll-to-Roll Transfer of Graphene Sheet

Roll-to-roll continuous synthesis and continuous transcription machinery based on the lab-scale success are being developed. Different types of roll-to-roll transfer apparatus and their set-up have been used by scientists. There are many apparatus being proposed and developed. It would be difficult to explain all of them. Some of them are enumerated here.

8.5.1 Patented Apparatus for Roll-to-Roll Graphene Synthesis and Transfer by the Research and Business Foundation at Sungkyunkwan University

(Inventors Byung Hee Hong, Jonghyun Ahn, Sukang Bae, Hyeong Keun Kim).

In this patented set-up and a published work by Bae *et al.* (2010), a graphene roll-to-roll transfer apparatus comprises five rollers.On the first roller unit, layers of graphene are synthesized on a flexible substrate. The second roller unit removes the substrate from the layered structure by immersing the layered structure provided by the first roller unit into an etching solution (it could be ammonium persulfate ($(NH_4)_2S_2O_8$), HF, BOE, $Fe(NO_3)_3$, iron(III) chloride ($FeCl_3$) or $CuCl_2$) and transfers the graphene layer onto the first flexible substrate at the same time. The third roller unit transfers the graphene layer from the first flexible substrate onto a second flexible substrate. Then a fourth roller unit forms a protection layer on the graphene layer formed on the first flexible substrate or the graphene layer

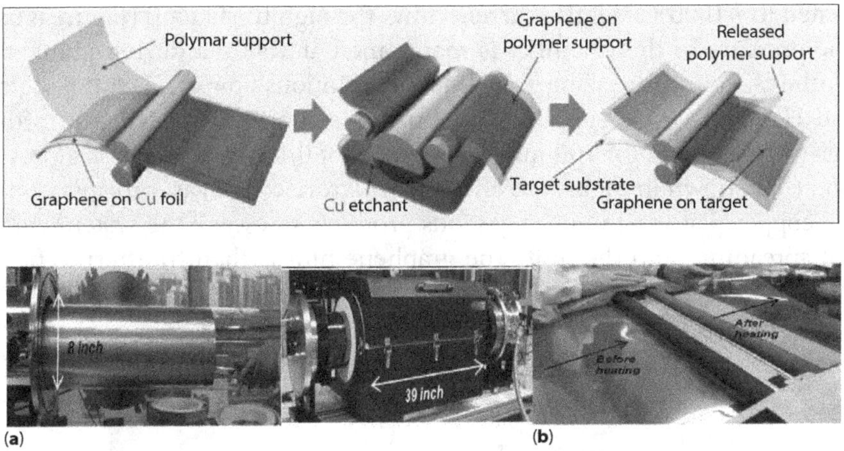

Figure 8.5 Upper three figures are schematic diagrams of the roll-based production of graphene films grown on a Cu foil, which includes adhesion of polymer supports, Cu etching (rinsing), and dry transfer-printing on a target substrate. A wet chemical doping can be carried out using the similar set-up used for etching. Lower three photographs are units used for the roll-based production of graphene films. (a) A Cu foil wrapping around a 7.5-inch quartz tube to be inserted into an 8-inch quartz reactor. The lower image shows the Cu foil reacting with CH_4 and H_2 gases at high temperatures. (b) Roll-to-roll transfer of graphene films from a thermal release tape to a PET film at 120°C. Copyright © 2010, Rights Managed by Nature Publishing Group, Courtesy: Bae *et al.*, *Nature Nanotechnology* 5, 574–578 (2010) doi:10.1038/nnano.2010.132.

formed on the second flexible substrate. Finally a fifth roller unit that is provided between the second roller unit and the third roller unit performs the cleaning and drying process to the layered structure. In Figure 8.5, only three basic specific rollers are shown.

This process appears to be useful for making large size graphene sheets. It may also be possible to control the deposition process (not show in Figure 8.5) to achieve the desired number of layers.

8.5.2 Four Roller Roll to Roll System

Kobayashi *et al* (2013) have described a different roll-to-roll system apparatus. The first set comprises of continuous roll-to-roll CVD system. The rollers are enclosed in a vacuum stainless steel chamber having a pair of winders, and a pair of current-feeding electrode rollers. All rollers in the system are electrically isolated from each other. One of the roller is coated with 230mm wide, >100m long, 36 μm thick, >99.9% pure Cu foil, which is placed between two current-feeding electrode rollers and by Joule heating, heated to ~1000°C. All the current flows through the Cu foil (Figure 8.6a) The pressure of this chamber is maintained at 1000 Pa during graphene synthesis. For synthesizing graphene, a continuous flow of CH_4 (450 sccm) and H_2 (50 sccm) is supplied and Cu foil is continuously wound on the roll-to-roll system for 100 min at a velocity of 0.1m/min. There is a provision (not shown in Figure) to allow the mixture of the gases to spray only the copper plate and unused gaseous products to escape the system without spreading onto the unit. The graphene film is then transferred from

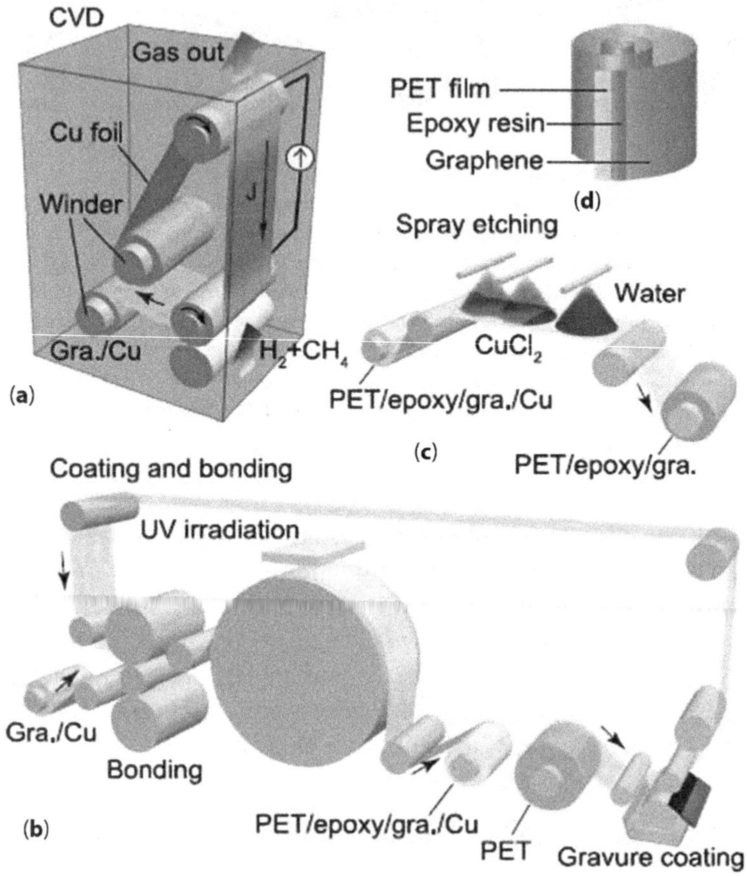

Figure 8.6 (a) Continuous roll-to-roll CVD system using selective Joule heating to heat a copper foil suspended between two current-feeding electrode rollers to 1000°C to grow graphene (Gra.). (b) Reverse gravure* coating of a photocurable epoxy resin onto a PET film and bonding to the graphene/copper foil, followed by curing of the epoxy resin. (c) Spray etching of the copper foil with a $CuCl_2$ solution. (d) Structure of the fabricated graphene/epoxy/PET film. Rights managed by AIP Publishing LLC. Courtesy: Kobayashi *et al.*, *Applied Physics Letters* 102, 023112 (2013); doi: 10.1063/1.4776707

the copper foil to a SiO_2/Si substrate using a poly [methyl methacrylate] (PMMA) as support film.

The next set (Figure 8.6b) comprises rollers for coating of a photocurable epoxy resin onto a PET film and bonding to the graphene/copper foil. Epoxy resin is then cured and Cu foil is etched away by spraying a copper chloride solution. Using this setup, all transfer processes can be performed using the roll-to-roll process. Since the rotation of roller containing the Cu takes almost 16 h, the time required for copper to be dissolved by etching agent has been nicely adjusted so that the chemical has enough time to react with copper to remove it from the polymer on which graphene is intact. By this method 100-m-long graphene transparent conductive film has been synthesized. This process seems to be latest method developed thus far.

8.5.3 Yamada's Method

Yamada *et al.* (2012) developed a roll-to-roll method for the graphene deposition machine using microwave plasma (MWP) CVD, (Figure 8.7) for industrial mass production. This is a continuous process producing 294-mm-wide graphene on a Cu foil. The process operates at 100–400°C temperature. Graphene film can be transferred onto a flexible sheet of Polythene tetraphthalate or quartz glass plate after Cu removal. The film shows high uniform optical transparency. Each quartz tube contains a coaxial linear antenna to generate plasma. These antennas are air-cooled. For graphene deposition, they used CH_4, H_2 and Ar, which are introduced into the CVD unit with flow rates of 30, 50 and 20 sccm respectively. The entire unit is maintained under vacuum. The roll-to-roll system consists of a pair of winder and unwinder as shown in Figure 8.7A and B. These

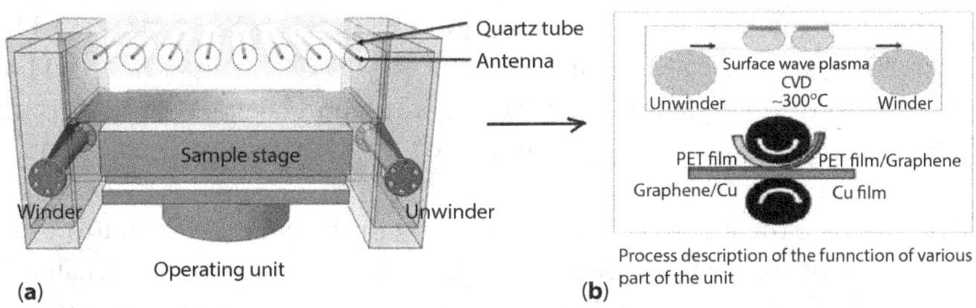

Figure 8.7 (a) Schematic of roll-to-roll graphene deposition machine and (b) Showing the function of each part of the unit. Copyright © 2012 Elsevier B.V. All rights reserved. Courtesy: Yamada *et al.*, *Carbon*, 50 (2012) 2615–2619.

winder/unwinder rotates with a speed which can be varied from 1–500m/s. Cu foil is exposed to the plasma.

8.6 Considerations for Minimizing Defects or Cracking During Transfer

Graphene, a two-dimensional material, is very susceptible to structural irregularities. Moreover, due to its topology defects (5-pentagons and 7-heptagons) it is often paired up into 5-7 dislocations and shows elevations.

Cracking or occurrence of defects in graphene films during transfer has been a major consideration for mass production. This can be avoided by ensuring good adhesion between the target substrate and the transferred graphene film so that tearing does not occur.

8.6.1 Selecting Proper Target Substrate

During the transfer of graphene from the substrate onto which it is synthesized, to a target substrate, graphene experiences chemical (from etchant), mechanical (removal by forcing its separation from one substrate to the other) or thermal forces. These forces disrupt the morphology and topography of graphene, causing cracking and thus inducing defects. It is important that target substrate have very good adhesion property. Hydrophobicity controls the adhesion of the graphene film; hence, a target substrate with higher roughness than the polymeric support would favor adhesion. Commonly used target substrates for graphene have included PET, Glass, SiO_2/Si (300nm thick) and also cleaved mica (Grade VI).

For adhesion of graphene onto the target substrate, it is crucial to have a very clean surface. That is why when SiO_2 is used as a target substrate, it is either sonicated in RCA SC1 solution (NH_4OH 28%: H_2O_2 30%: H_2O = 1: 1: 5) at an increased temperature of 50–55°C (Liang *et al.* 2011), or treated with oxygen plasma for cleaning. Oxygen plasma treatment of the target substrate improves the adhesion of the graphene and also reduces the chances of forming cracks.

To enhance the adhesion of graphene by limiting crack formation during etching of the PMMA, the target substrate needs to be pre-treated by hydrophobic self assembled monolayers (SAM), perfluorophenylazide (PFPA) and aminopropyl-triethoxysilane (APTES). Though the use of PMAA for coating graphene prior to transfer for production of continuous graphene is one of the most suited methods, transferring graphene using PMMA requires advanced knowledge including wet chemistry and very careful handling.

Another alternative is to reduce the roughness of the substrate onto which graphene is synthesized.

8.6.2 To Avoid the Use of Etchants

Theoretically, transfer by elastomeric stamps such as PDMS can avoid exposure of the graphene film to liquid etchants and the need to utilize a polymeric support. But so far, PDMS has not revealed remarkable improvements in graphene quality. There is a need to look for an alternative to PDMS or to develop a methodology using PDMS that can produce better quality graphene.

Another way is to transfer graphene is using thermal release tape. In this case, first the protective layer should be peeled off from the thermal release tape, and then the adhesive layer of the tape should be put on top of the graphene layer. It is imperative that the pressure applied to the tape should be standardized to avoid cracking. Usually firm pressure to the tape is applied using a rubber roll. There are reports of using a spatula to apply pressure; however, we feel that rubber rolls are better.

It has been observed that obtaining continuous graphene when thermal release tapes are used for transfer is very difficult.

In the end, we would like to mention that the perfection in producing graphene is difficult to achieve. However, there are some defects that are required or even induced for a particular application or for inducing specific property in graphene. Some examples are listed here.

i. Magnetism is induced by point defects and edges in graphene. Yazyev and Helm (2007) and Yazyev (2008) have shown that in graphene, the single-atom defects (e.g., vacancies and hydrogen chemisorption) induce spin-polarized defect states. According to Yazyev, coupling between the magnetic moments is either ferromagnetic or anti-ferromagnetic, depending on whether the defects correspond to the same or to different sub-lattices of the graphene lattice, respectively. The lattice distortions near the GB change the electronic structure in its vicinity, which causes scattering of electrons thus affecting the transport phenomena both across and along the GB direction.

ii. Imperfections in graphene have been found to improve the performance of graphene (Salehi-Khojin and Estrada 2012). According to them, "nearly-pristine graphene chemiresistors are less sensitive to analyte molecules because adsorbate bind to point defects, which have low resistance pathways around them."
iii. Yazyev and Louie (2009) have constructed and shown that dislocations and grain boundaries in graphene, extended defects affects the electronic structure and transport properties.
iv. Imperfections that disrupt the sp^2 conjugation of graphene can alter its electrical, chemical, and mechanical properties. According to Tan *et al.* (2012), graphene defects have higher reactivity.

8.7 Summary

In this chapter we have presented the development of various techniques used for developing commercially viable nanoimprint lithography method in chronological order (Table 8.1). Nanoimprint lithography is a method of fabricating nanometer scale patterns by nanolithography process; this method is used because it is a low-cost, high-throughput and high-resolution method. The efforts made can be categorized as (i) plate-to-plate and (ii) roller-based methods. As it can be seen in Table 8.1, first plate-to-plate production of graphene was attempted in 1995, and then roller-based graphene synthesis and transfer method evolved in 1998. Both rigid and flexible synthesis substrate as well as target substrate have been used in roll-to-roll synthesis and transfer.

Table 8.1 Nanolithography methods used for commercial production of graphene.

Nanolithographic Methods for Synthesis and Transfer of Graphene *Using Plates*		
Plate to Plate	*Thermal*	• Chou *et al.* 1995 • Chou *et al.* 1996 • Lenin *et al.* 2002 • Perret *et al.* 2004 • Chen *et al.* 2006 • Merino *et al.* 2007 • Park and Cheng 2009
	UV	• Volger *et al.* 2007 • Shinohara *et al.* 2008 • Mohammed 2009 • Ye *et al.* 2010
Nanolithographic Methods for Synthesis and Transfer of Graphene *Using Rollers*		
Roll to Plate (on rigid substrate)	*Thermal*	• Tan *et al.* 1998 • Song *et al.* 2010 • Lan *et al.* 2010 • Lim *et al.* 2011
	UV	• Kim *et al.* 2009
Roll to Plate (on flexible substrate)	*Thermal*	• Tan *et al.* 1998 • Youn *et al.* 2008
	UV	• Lee *et al.* 2008 • Park *et al.* 2009
Roll to Roll (on flexible substrate)	*Thermal*	• Makela *et al.* 2007 • Nagato *et al.* 2010 • Makela *et al.* 2011 • Schleunitz *et al.* 2011
	UV	• Ahn and Guo 2008 • Hwang *et al.* 2009 • Makela *et al.* 2010 • Maury *et al.* 2011 • Unno and Taniguci 2011

9

Graphene in Industry, Commercialization Challenges, and Economics

Advances in technology will continue to reach far into every sector of our economy. Future job and economic growth in industry, defense, transportation, agriculture, health care, and life sciences is directly related to scientific advancement.

Christopher "Kit" Bond, former U.S. senator

9.1 Introduction

In this chapter, various aspects of commercialization challenges, both technical and economy-related issues of graphene and graphene-based products are discussed.

It has been on the agenda of many scientific bodies and governments to take graphene-based products from a state of raw potential to a point where they can revolutionize the industries so as to bring a new dimension to future technology, producing faster, thinner, stronger, flexible, and broadband products of graphene that can give a high return on investment.

The economy advances along with technology. For better chances of success, today's economy needs more universal reliable technology. The question is: Is graphene the universal technology material to revolutionize today's industries? We get positive answers to this question when we look at the plethora of outstanding patents being filed. However, the big question remains of how many applications or patents will be used and will live up to the expectations. Fortunately, at the present time, the phenomenal research and development activity making graphene suitable for industrial applications are very encouraging towards the development of a graphene industry in the coming years.

Specific properties, like high electrical conductivity, high mobility for electrons as well as for holes, high thermal conductivity, high Young's Modulus and dependence of these properties on a number of graphene layers are the some of graphene's major assets, which have attracted scientists and technologists to develop commercially viable large-scale production of large-size graphene sheets. Let us have a quick look at the graphene-based products that are already on the verge of entering into the market, as this will give insight into the importance of graphene in emerging industries. Opportunities in the near future are in power management (graphene for harnessing solar energy), hybrid electronics/flexible electronics (utilizing graphene's high electrical conductivity), transparent conducting film (with a view to replace conducting glasses), displays energy/supercapacitors (storing large DC power for more than 300–400F/g), photovoltaics (development of conducting transparent electrodes for dye-sensitized and organic flexible solar cells), thermal management (utilizing graphene's high thermal conductivity for using it as a heat sink in electronic circuits), etc.

Realizing graphene's potential, governments, industries and private investors are funding graphene research so as to accelerate the developments of industrially viable products for its commercialization. Considering various applications of graphene, the greatest challenge is in mass production of high quality graphene with the desired layers and production of graphene-based material.

If patents are an indicator of commercial growth, graphene is not lagging behind as graphene synthesis has attracted a lot of attention since 2008. According to Thompson Innovation database, USA tops the list for the highest number of patents with 31.76%, followed by South Korea (19.20%), China (10.35%), Taiwan (2.58%), Germany (1.88%), France (0.70%), Australia (0.47%), Canada (0.47%), Great Britain (0.70%) and others (1.17%). One should not be disheartened by seeing a lesser number of patents by any specific country, because it is also equally important to see the statistics of how many of these patents have really been taken up by industrialists for mass production. Since we could not obtain this information, it is not possible to give this figure. There is no doubt that the number of patents is an indicator of the interest and importance of this field globally. So far as the worldwide assignees of the patents are concerned, Thompson Innovation database analysis has revealed that academic/research institutes along with independent investigators are the leaders, with about 50% of patents. Among the industries, Samsung Group is the leader. Other leading companies are Guardian Industries Corp., Hitachi Ltd., Nanotek Instruments, Hewlett Packard Development Co., L.P., Fujitsu Ltd., IBM Corp, and others.

Many collaborative efforts by industries and institutes or universities to translate the potential application of graphene into electronic devices, advanced batteries, supercapacitors, multifunctional composites and so on, are underway. The maximum collaboration is in South Korea and the USA, and the most active company is Samsung Electronic Corp. It is collaborating with Sungkyunkwan University (SKKU), Hanyang University and Leyland Stanford Jr. University, USA.

Interest in graphene started in 2004 and it has only been 10 years during which scientists, technologists and industrialists have indulged in making graphene as a possible industrial product. Graphene has to go a long way before it will find its commercial products in the market. Nevertheless, examining the topics of research being published in international journals and the nature of patents, it may not be a wrong assessment that graphene will help finally in miniaturizing electronic circuits, thus making electronic gadgets even smaller.

9.2 Graphene Industries

Like its parent body, graphite, graphene has also made a special place in industry. There are ~ 106 graphene companies around the world, involved in research and development and production. Most are in North America. Companies working on graphene can be categorized as follows:

1. Companies producing graphene
2. Companies developing graphene-based products
3. Graphite mining companies
4. Companies making equipment for manufacturing graphene

5. Graphene related companies that provide software, technologies and other services for graphene

Some of these companies working in the graphene sector are given in Tables 9.1 to 9.5.

9.2.1 Companies Producing Graphene and Graphene-based Applications

A series of development for the manufacturing of graphene sheets have been completed (as discussed in chapters 7 and 8), but the roll-to-roll method for the production of large size graphene on a large scale seems commercially viable. However, the stock investors, though seeing great potential in graphene industry, are baffled at the moment about investing in graphene-based industries. This has slowed down industrial development. The reasons are that the roll-to roll production processes have yet to move out of the lab and the quality and size of graphene sheets produced in this way is still being upgraded.

Lockheed Martin, a global security and aerospace company, has obtained a patent for perforated graphene nanopore-based perforene water filters for testing it in a desalination plant as well as for managing the oil and gas industry's wastewater. These graphene sheets have precisely sized 1 nm holes. They have started producing graphene, but the company has not yet commercialized the technology. Some mutual funds are still researching investment in graphene-based technology. Garvin Jabusch and Frank Morris have cautions about the graphene industry:

> Widespread commercial viability of graphene's properties may still be further off than many investors seem to be hoping. This is mainly because a lot of its benefits are paired with limitations that to greater or lesser degrees, still need to be overcome. For example, even though it has 100% energy conversion rate for absorbed photons, it only absorbs ~3% of photons striking it. There is a lot of research going on addressing this such as with dye-sensitized cells and other ideas, but all of this is primarily still in the lab and not ready for widespread use. Keep in mind, graphene was only discovered in 2004. Something we've been aware of for less than a decade will necessarily have a long way to go in terms of our understanding how to best unlock its potential (Forbes Investing,18th September 2013).

Statements made by Garvin Jabusch and Frank Morris may be true for the specific applications they have mentioned, but any developments based on graphene's electrical and thermal property can find commercial application soon, especially in miniaturizing electronic circuits and thus making electronic gadgets smaller than what is available today. We need to take an optimistic view of graphene rather than a pessimistic one.

There are very few companies around the world who are commercially producing graphene. Many company's efforts are focussed on standardizing the use of graphene for a particular application and then, accordingly, manufacturing the graphene. For example, Graphenea, Bluestone Global Tech, 2-D Tech, and Graphene Frontiers, are developing novel manufacturing technology to allow them to manufacture high quality single-layer graphene sheets for their application in higher value, longer-term prospects like touchscreen displays and improved photovoltaic cells.

Recently, a Canada-based company, NanoXplore, has gone in full swing to produce three tons of pristine graphene GNPs per year from graphite. It has the largest graphene production capacity in Canada and claims to be the fifth largest in the world, excluding China. Moreover, it has perfected a process that can functionalize graphene during production. This company is going ahead by raising $2.18 million for its sales effort.

Table 9.1 Some graphene-producing companies.

S.N.	COMPANIES	PRODUCTS
1	**ACS Material** USA based company	CVD grown single-layer graphene, graphene oxide, graphene nanoplates, carboxyl graphene and graphite oxide
2	**Anderlab Technologies** Based in Mumbai, India	Claim they can manufacture graphene on a ton scale. They are also developing graphene inks and nanocomposites
3	**Angstron Materials** Based in Ohio, USA	Nanographene platelets, pristine graphite and single layer graphene
4	**Applied graphene Materials** A spin-off from Durham University, UK	Graphene synthesized by patented scalable "bottom-up" CVD approach
5	**Bluestone Global Tech** Based in New York, USA	High-quality, fully customizable graphene on quartz, copper, silicon and other substrates
6	**Cambridge Nanosystems** A spin-off from the University of Cambridge UK	Graphene of high purity

(*Continues*)

Table 9.1 (Cont.)

S.N.	COMPANIES	PRODUCTS
7	**CTI Nanotechnologies** In collaboration with American graphote Technologies	Producing graphene paper
8	**CVD Materials Corporation** A subsidiary of CVD Equipment	Monolayer graphene, multi layer graphene and 3D graphene
9	**Directa Plus** An Italian company	Super-expanded graphite, pristine GNPs, water-dispersed GNPs and fine nanographite powder all produced by exfoliation process
10	**Future Technologies** Based in UK by Sri-Lanka-based GS International and RS Group	Graphene
11	**Garmor Inc** A spin-off from University of Central Florida based in Orlando, Florida, USA	Graphene and graphene oxide flakes
12	**Grafen Chemical Industries** based in Turkey	Graphene of excellent quality
13	**Grafoid** A private company based in Canada	Graphene
14	**GRAnPH Nanotech** Based in Spain	Single-layer graphene, based on research by Grupo Antolin-Ingenieria and the Chemical Engineering Department of the University of Alicante in Spain
15	**Graphenea** A private European company based in Spain	High quality single-layer, bi-layer and multi-layer graphene sheets, graphene oxide on any substrate that the customer provides.
16	**Graphene Devices** Based in Niagara Falls, USA	Graphene using a process developed at the University of Buffalo

(*Continues*)

Table 9.1 (Cont.)

S.N.	COMPANIES	PRODUCTS
17	**Graphene Frontiers** Based in Pennsylvania, USA	Graphene synthesized by "Atmospheric Pressure CVD" roll-to-roll room-pressure process. Technology was developed at the University of Pennsylvania
18	**Graphene Industries** Based on outputs of Professor Andre Geim's research group at Manchester University in the UK	Supplies graphene for micro- and nano-electronic fabrication and research. The company supplies graphene materials to MU's 2-DTech
19	**Graphene Laboratories** Canada based	Functional graphene materials and devices. Joint venture company with Lomiko Metals.
20	**Graphene Technologies** Based in Novato, California, USA	Produces graphene by a unique patented and eco-friendly method from carbon dioxide
21	**Graphensic** Based in Sweden, a spin-off from the Linköping University	Produces epitaxial single layer graphene on hexagonal SiC by high temperature process for the electronic equipments market.
22	**Graphite Zero** Is a spin-off of the National University of Singapore	Produces graphene using Focus Metal's high quality graphite.
23	**Haydale** A UK based subsidiary of the ICL Group	Supplies Split Plasma treated graphene flakes, graphene-based inks and R & D services
24	**Incubation Alliance** Based in Kobe, Japan	Graphene Flower materials: substrate-free and catalyst-free multi layer graphene material made by direct synthesis
25	**KNano** full name is Xiamen KNano graphene Technology Co	Several types of graphene platelets and polymer-graphene composites.
26	**Morgan Advanced Materials** Based in the UK	Scaling-up a new graphene manufacturing process along with with the University of Manchester

(*Continues*)

Table 9.1 (Cont.)

S.N.	COMPANIES	PRODUCTS
27	**Nano Carbon** Poland; is co-owned by mining giant KGHM and the Industrial Development Agency	Graphene using technology developed at the Institute of Electronic Materials Technology in Warsaw
28	**NanoCarbon Pty Limited** Established in 2014 in Australia	Pilot production of surfactant-free graphene (developed at the University of Wollongong) by July 2015 and also graphene based high barrier films, lithium ion batteries and water purification
29	**National Nanomaterials** USA-based company in collaboration with Texas State University	Graphene Functionalized with amine, amide, ester, carboxylic, or hydroxyl functional groups
30	**Ningbo Morsh Technology** Based at Ningo, Zhejiang.	Graphene using technology developed at the Chongqing Institute
31	**Perpetuus Carbon** Based in the UK	Graphene production capacity of 100 tons to be extended to 500 tons
32	**Saivens** An Indian Company	High purity (over 90%) graphene sheet grown using CVD
33	**Shanghai SIMBATT Energy Technology** Shanghai-based China	Graphene powder, intercalated graphite, graphene oxide, doped graphene powder, graphene quantum dot and 1-5 layer graphene powder
34	**The Sixth Element Materials** Based in Chengzhou, China	Graphene flakes, graphene-powder and graphene oxide. Production capacity is 100 tons/year, to be increased to 1,000 tons by 2016
35	**XG Sciences** Private company based in Michigan, USA	Composite Materials supported by the technology from the Michigan State University College of Engineering
36	**Xolve** US-based company	Graphene and nanographene
37	**NanoXplore** Canada-based group	Production of 3-ton GNP from graphite

Once the graphene is synthesized, the next step is to process the graphene into a component, such as an electronic transistor, for mobile phones or a bulletproof vest, and so on. Products like bulletproof vests need a mixture of graphene with something else, such that graphene retains its particular desired quality for a specific use. Most of these products are produced and patented by small companies, which are ready to transfer their technologies to larger companies for production and marketing to the end user. At the moment there are only two companies, AMO and Bluestone Global Tech, that are manufacturing and selling components for modern technologies, such as graphene transistors, photodetectors, i.e., end-user products. There are many companies such as IBM, Samsung, and Sandisk, which are involved in the graphene production to application and end-user demand. Such companies are mostly using CVD graphene.

In March 2015, a Chinese company, Moxi, along with a device-maker company, Galapad, has announced the first mass production of a graphene-enhanced phone named *Galapad Settler*, using 5.5" graphene as touchscreen. The company claims that this touchscreen is more sensitive and produces better color than conventional touchscreens. Moreover, they have used graphene in the battery, which raises the charge density by 10% and extends the battery life by 50%.

Graphene Frontier is a spin off from the University of Pennsylvania. They are developing the Atmospheric Pressure CVD roll-to-roll process. Backed with a funding of $1.6 million, Graphene Frontier is manufacturing a six sensors brand for highly sensitive chemical and biological GFET-based sensors to be used in disease diagnosis with multiple markers such as cancer.

The UK and European governments are backing the development of commercial facilities and funding further exploration of the ways in which graphene's applications can be exploited.

Graphenea in collaboration with Philips and the University of Cambridge have developed a monochrome OLED device using molybdenum trioxide (MoO_3) film as an intermediate layer between the graphene and the OLED layers as the transparent conductor layer. This graphene-based TC outperforms ITO, which at present is the most popular material for transparent conductors in displays and solar cells, but it is expensive, rare and brittle.

Companies like Haydale and Vorbeck Materials are working on exploiting the market for the applications of graphene flakes and graphene nanoplatelets to create advanced composites and conductive inks for printable electronics.

A few more companies are involved in graphene 3D-printing efforts, namely Grafoid, Graphene 3D Labs, AGT, Kibaran, Qingdao Unique Products, Graphene Technologies and Stratasys.

Stratasys is a leading 3D-printing company, worth over $5 billion, and this effort will be partly funded by the US-Israeli BIRD Foundation.

Microdrop Technologies (a Germany-based company) is using graphene provided by Talga Resources and have tested it for applications in micro printing, 3D printing and other related applications.

Samsung Advanced Institute of Technology (SAIT), through its partnership with Sungkyunkwan University's School of Advanced Materials Science and Engineering, became the first in the world to develop a new method to produce large area, single crystal wafer scale graphene on a semiconductor, maintaining its electric and mechanical properties. This single-crystal monolayer of graphene is wrinkle-free. It is grown on silicon wafer, using a germanium buffer that is devoid of hydrogen. The advantage of this system is due to the anisotropic two-fold symmetry of the germanium (110) surface. Many seeds were formed on this surface, and these seeds merged to form single-crystal graphene. Another advantage of a hydrogen-terminated germanium surface is its weak interaction with graphene that allowed etch-free dry transfer of graphene, leaving geranium to be recycled. Some other such innovative companies that are developing graphene-based applications are listed in Table 9.2.

Table 9.2 Companies developing graphene-based applications.

S.N.	COMPANIES	PRODUCTS
1	**Alphagraphene**	Chips for transistors using graphene FET on Si Wafer. Each chip consists of an array of 36 graphene FETs with nine different graphene channel length/ width arrangements to accommodate all research and development settings. The high-mobility CVD graphene is used as the transistor channel
2	**Apex Graphene** USA company	i. Transparent conductive film ii. Heat spreader sheet, both of up to 15 meters in length
3	**CalBattery** Los Angeles, USA-based company	Silicon-graphene lithium-ion battery anode material

(*Continues*)

Table 9.2 (Cont.)

S.N.	COMPANIES	PRODUCTS
4	**Calevia** Canadian Company established by Grafoid and ProScan Rx Pharma	To develop graphene-based cancer thermal treatment platform, to overcome the side effects and strong limitations of common cancer therapies
5	**Cientifica** UK-based company	To establish business using graphene
6	**Firmus SAM** Based in Monaco with production facility in Warsaw, Poland	Water treatment membrane using graphene
7	**Graphenea** Private European company based in Spain	High quality single-layer, bilayer and multilayer graphene sheets and graphene oxide for industrial applications
8	**Graphene 3D Lab Inc.** A spin-off from graphene Labs	Development of high-performance graphene-enhanced materials for 3D printing
9	**3D Graphtech Industries** Established in 2014 by Kibaran Resources and 3D Group	R & D for graphite and graphene applications for 3D printing. They source their graphite exclusively from Kibaran's Tanzanian graphite mine
10	**Graphene Batteries** based in Norway	Is developing safe and durable graphene based high-energy battery materials
11	**Graphene Energy**	Developing graphene for electrode for next generation nano-technology based ultra-capacitors for energy storage that will have at least twice the storage capacity of commercially available ultra-capacitors
12	**Graphenhex** based in the UK	R & D for high-energy supercapacitors based on graphene materials
13	**mPhase Technologies**	Its flagship product (not yet released) is the AlwaysReady Smart NanoBattery—which offers reliability, extended life, and safe disposal. mPhase is exploring the printing of its "Smart NanoBattery" using graphene

(*Continues*)

Table 9.2 (Cont.)

S.N.	COMPANIES	PRODUCTS
14	**SiNode Systems**	To commercially produce a novel anode Li-ion battery developed at Northwestern University. The anode of this battery uses a composite material of Si nano-particles and graphene in a layered structure
15	**Vorbeck Materials** a USA company	Graphene based inks for the printed electronics (Vor-X). Plastic and rubber composites with low Vor-x loadings attain extreme levels of strength, dimensional stability, conductivity, and environmental resistance, opening new application and design possibilities
16	**Vulvox Nanobiotechnology Corporation** Based in Long Island, NY	R & D for manufacturing ultra high strength graphene for ballistic armor and structural materials applications and also for applications in power generation, electric motors and high temperature materials processing

Big electronics companies are concentrating their research efforts to integrate graphene into their future electronic products. In 2014 Samsung Electronics announced a breakthrough synthesis method to speed the commercialization of graphene, a unique material ideally suited for electronic devices. According to a recent report in *Graphene News*, IBM has announced (in July 2014) to invest $3 billion to find next-gen chip technology. This next-gen chip technology is mainly to meet the growing demand by cloud computing and big data systems. Their efforts will be directed in two stages:

In the first stage they will be utilizing existing silicon technology to fabricate transistors in a 7 nm process.

In the second stage they will be focusing on developing an alternative to silicon technology because of the physical limitations of silicon-based semiconductors. Nowadays, bandwidth to memory, high-speed communication, and device power consumption are the demand to be met with higher efficiency and orders of magnitude improvement in system level performance and energy efficient computing. Apart from graphene, they are also looking into CNT, quantum computing, silicon photonics, and so on.

Another unique application of graphene is in a graphene-enhanced drilling fluid, named *PlatDrill*, which is being manufactured by Graphene Nanochem. They are to deliver 30,000 tons/annum of PlatDril worth $1.7 million. Graphene-enhanced PlatDrill, is drilling fluid that aims to deliver superior performance with lower environmental impact as well as better lubricity, load bearing capacity and higher viscosity properties.

Structured Materials Industries (USA), in collaboration with the Department of Electrical and Computer Engineering, Cornell University, are producing deposition tools and support equipment for graphene manufacturing. The tools manufactured by them are:

i. Horizontal desk tube reactor (CVD, PECVD graphene/CNT Metal substrates) that has a heating capacity of up to 1200°C. This is suitable as R & D reactor.
ii. Vertical RDR Tool (1700°C) that along with CVD can possibly be used for wafer growth and doping.
iii. Multizone horizontal tube suitable for CVD, TD, SiC CVD wafer growth (1200°C, 1700°C or 2200°C), and
iv. Multizone tape tool for roll-to-roll CVD on metal tape for continuous graphene synthesis. It can be used for foil preparation and post-process doping, etc.

Recently, Focus Metals Inc., a Canadian mining company in partnership with Grafoid Inc. (established in September 2011) have started a pilot graphene plant using graphite. They have started production of high-purity and high-energy density graphene in August 2014.

With the manufacturing side of the graphene industry in clear growth, translating it to applications is more important than ever to maintain the momentum. All in all, graphene's future still looks very bright.

XG Sciences Inc. is the manufacturer of graphene nanoplatelets and engineered materials based on nanoplatelets. Moreover, they also do customer-oriented work for specific applications. They are collaborating with Oak Ridge National Laboratories to develop graphene titanium composite as an important future material for many applications in industrial and military sector.

IP Group of UK's (an intellectual property commercialization company) portfolio company, Applied Graphene Materials, has announced its intention to enter into graphene-based products and wishes to raise funds. It has a partnership with many universities to develop innovative graphene-based products.

The trend of industries working in collaboration with universities and research institutions is very prevalent in developing graphene-based products.

9.2.2 Companies Supporting Graphene-related Activities

The world reserve of graphite has been estimated to be 800 million metric tons. China produces 75% of the world's graphite. The second largest reserve of graphite is in Indian mines, which contain 11 million tons of amorphous graphite.

9.2.2.1 Graphite Mining Companies

Graphite, being one of the raw materials for the synthesis of graphene, graphene oxide and graphene nanoribbon, is a necessity for graphene synthesis. Natural graphite is mined either from open pits or underground. China is the biggest exporter of natural graphite followed by India, Brazil, North Korea and Canada. Australia and Sweden also have graphite mines. Synthetic graphite is mostly produced in the United States from carborundum by heating it a very high temperature to remove the silicon. Though synthetically produced graphite is of the highest-quality, known as Highly Oriented Pyrolytic Graphite (HOPG), compared to natural graphite, synthetic graphite is very costly. Hence, most of the commercial graphene-producing companies are collaborating with graphite mining companies. Graphite is consumed in many industries such as refractories, batteries, steelmaking, expanded graphite, brake linings, foundry facings, lubricants, as a matrix and neutron moderator within nuclear reactors and as radar absorbent materials, and presently companies prefer to use CVD method for the mass production of graphene.

There are many graphite mining companies that are supporting research and development and manufacturing of graphene and graphene-related application products. Some of them are listed in Table 9.3. Graphite mines in China, Sri Lanka, Sweden, Canada and Australia are playing an active role in graphene industries. China has more than half the reserves of graphite and has realized the technological importance of graphite. To protect its resources, it has raised the export taxes on graphite. Others believe that this action is unwarranted because graphene consists of only one layer of carbon, therefore, a single graphite mine would be enough to supply the world with graphene even if all the usage of graphene materializes. Moreover, nowadays most large-area graphene is grown with the CVD process, which may not have superb quality, but is suitable for most of the purposes.

Table 9.3 Graphite mining companies.

S.N.	COMPANIES	PRODUCTS
1	**American Graphite Technologies**	A mineral exploration and technology development company that has 100% ownership of 100 mineral claims in Quebec, Canada. It is also involved with graphene related research
2	**China Carbon Graphite Group** based in Chengguan Town (Xinghe), China	Manufactures and sells graphite-based products (graphite electrodes, fine grain graphite blocks and high purity graphite) in China
3	**Focus Graphite** based in Ottawa, Canada	A mine developer and the owner of the highest-grade (roughly 16%) technology graphite resource in the world (at Lac Knife in Quebec)
4	**Lomiko Metals Inc.** Canada-based company	It is an exploration-stage company engaged in the acquisition and development of properties that contain minerals for the new green economy. Its mineral properties include the Quatre Milles graphite Property and the Vines Lake property.
5	**Mason Graphite** Canadian mining company	Exploration and development of its 100% owned Lac Gueret graphite project in northeastern Quebec
6	**Mega Graphite** based in Ontario, Canada	It mines, process, purifies and supplies natural graphite. The company holds four graphite mines in Canada and the Uley graphite Mine in South Australia
7	**National Graphite Corporation** based in the USA	Is focused on bringing the Chedic graphite Mine back into commercial production
8	**Northern Graphite** Canadian mine devel-opment company	Its main asset is the Bissett Creek graphite project located 100 km east of North Bay, Ontario. The company is also involved with graphene research

(*Continues*)

Table 9.3 (Cont.)

S.N.	COMPANIES	PRODUCTS
9	**RS Mines** Based in Sri Lanka	RS Mines (previously GS International and the RS Group), holds several 99.99% natural high purity crystalline vein graphite mines in Sri Lanka—including their flagship mine, the Queen's Mine. The company is also producing defect-free graphene and graphene oxide products
10	**Talga Resources** Australia-based miner company with gold assets in Australia and high-grade graphite resource in Sweden	Produces small-scale graphene products and aims to start a pilot plant in Sweden
11	**KS Mines** Palamu, Jharkhand, India	The only mine in India that has 1.8 million metric tons of flaky graphite

9.2.2.2 Companies Making Graphene Manufacturing Equipment

In Table 9.4 there is a list of some of the companies that make equipment for graphene synthesis. There are many more companies that are involved in the graphene and can be found by browsing the Internet. As we know, graphene is synthesized from graphite by liquid-phase exfoliation or another chemical "exfoliation" procedure. These procedures do not need very specific equipments.

Another process to synthetically make graphene is by chemical vapor deposition (CVD) of graphene onto a substrate. This method does not require graphite ore; rather it uses various gases as precursors. Graphene produced by CVD method is preferred by microchip and display manufacturers. CVD Equipment of New York and Aixtron of Germany are the leading manufacturers of CVD equipments such as tube furnace and other furnaces. There are many small companies that supply the substrate material for deposition and transfer of graphene to target substrate.

Table 9.4 Companies making graphene manufacturing equipment.

S.N.	COMPANIES	PRODUCTS
1	**Aixtron**	Provides deposition equipment to the semiconductor industry. The company's technology solutions are used to build advanced components for electronic and opto-electronic applications
2	**CVD Equipment** Based in Long Island, NY, USA	Offers a variety of process and support equipment, for both R & D and production facilities
3	**Moorfield** A UK based company	It designs and manufactures a wide range of R & D-scale laboratory systems, vacuum deposition products and CDV synthesis units
4	**MTI Corporation** Based in California, USA	Leading manufacturer of oxide crystals and substrates, lab equipment and furnaces and tube furnaces for graphene growth
5	**PlanarTech** USA-based company	Offers economical process equipment, analytical equipment and training services for emerging 2D materials (graphene, boron nitride, molybdenum disulfide, etc)
6	**Veeco**	Makes process equipment and molecular beam epitaxy (MBE) products, which can be used for graphene deposition
7	**VG Scienta** A merger between Vacuum Generators and Gammadata Scienta	For graphene (and other 2D materials), the company offers CVD deposition systems and also PVD and sputtering systems based on metal evaporation using thermal cells or E beam

9.2.2.3 Companies Providing Software, Technology or other Services for Graphene Industries

Abalonyx is a Norwegian company based in Oslo and it specializes in the synthesis and manipulation of graphene oxide and other graphene derivatives. With an aim to secure quality and reproducibility they use and develop automated, high throughput methods based on the standard well-plate format so as to get full automation of all experiments and production.

BASF a leading chemical company that has collaborated with the National University of Singapore to develop the use of graphene in organic electronic devices, such as organic light emitting diodes (OLED). The goal of this collaboration is to interface graphene films with organic electronic materials for the creation of more efficient and more flexible lighting devices. BASF, in collaboration with Max Planck Institute for Polymer Research, have initiated a Carbon Materials Innovation Center (CMIC), to research the scientific principles and potential applications of innovative carbonized materials and evaluating their potential uses in energy and electronic applications. The total investment for the joint research and development platform amounts to €10 million.

There are many companies that are providing software and helping in web designing to promote the R & D and production work in the area of graphene and its application (see Table 9.5).

Table 9.5 Graphene-related companies that provide Software, technology or other services for graphene industries.

S.N.	COMPANIES	PRODUCTS
1	**Abalonyx** Norway	Develops functional nano materials (nano-composites, nano-laminates and coatings) based on graphene derivatives. The company's main focus is functional coatings for the renewal energy sector
2	**Archimedes Polymer Technologies** Cyprus	A specialist developer of nano composite and graphene nanoplatelets
3	**Avansa Technology and Services**	Offers analytical testing and consultancy services for industries with emerging nano and micro technology based products. The company also provides wide range of nanotechnology based research services and products. The company offers direct material sales, and they supply few-layer graphene, functionalized graphene, reduced graphene oxide and also graphene in solvents
4	**BASF** Based in Germany	One of the world's largest chemical companies. BASF has been involved with graphene research since 2008 in collaboration with the Max Planck Institute

(*Continues*)

Table 9.5 (Cont.)

S.N.	COMPANIES	PRODUCTS
5	**Cambridge Graphene Platform** A spin-off from Cambridge University UK	Is to commercialize graphene inks based on research work done at Cambridge University
6	**CrayoNano** A spin-off from the Norwegian University of Science and Technology	To commercialize a new technology to grow gallium arsenide (GaAs) nanowires on graphene using molecular beam epitaxy. This new hybrid electrode material offers excellent optoelectronic properties
7	**CTDAT** A private company based in Mexico and in collaboration with Arizona University, UNAM and UNISON	A consultancy firm that specializes in graphene production technologies, have developed a process to exfoliate graphene from graphite, is able to produce graphene oxide and reduced graphene oxide from ore mineral graphite
8	**Future Carbon** Based in Germany	Develops and manufactures carbon nanomaterials and carbon supercomposites and has acquired several graphene and CNT related patents from Bayer
9	**Gnanomat** Based in Madrid, Spain	Developed a proprietary patented liquid-phase procedure to exfoliate graphite to graphene. The single-step process produces high-quality graphene micro platelets. The process uses cheap, biodegradable solvents
10	**GrafTech International** A global company	Has more than 120 years of experience in the carbon and graphite industry. Makes an expanded natural graphite foil that keeps most of the properties of the graphene (thermal, electrical, and possibly acoustic) and they manufacture them in flexible sheets from 20um to 1mm thick and in widths up to about 50"

(*Continues*)

Table 9.5 (Cont.)

S.N.	COMPANIES	PRODUCTS
11	**Graphene-Info** Based in Israel	It is a web publication focused on graphene material technologies. Graphene-Info has been providing news and resources for the graphene industry and researchers since 2009
12	**IBM (International Business Machines Corporation)** Based in the USA	It is a multinational technology and consulting corporation. IBM operates several research labs around the world and they are researching graphene related technologies—mostly graphene based transistors and photo detectors
13	**IDTechEx**	Offers independent market research, business intelligence and advice on graphene and also organizes global graphene events
14	**Intel Corporation,** Based in the USA	Intel is the world's largest semiconductor chip maker (based on revenue). The company supplies chips for PCs, mobile devices and more. Intel is researching graphene to be used in future electronic devices
15	**LG Corporation,** based in South Korea	LG is a multinational conglomerate corporation, involved with electronics, displays (LCD, plasma and OLEDs), telecom, chemicals and more. LG has an active graphene research and has filed for many graphene patents
16	**NanoIntegris**	It supplies high mobility semiconductor inks based on graphene nanoplatelets
17	**NanoXplore** Canada	It provides graphene services and products
18	**Nokia** Based in Finland	It is a large multinational corporation and was once the world's leading mobile phone maker. Nokia Research Center, with its 10 laboratories world-wide, is exploring new technologies mostly for mobility applications

(Continues)

Table 9.5 (Cont.)

S.N.	COMPANIES	PRODUCTS
19	**Oxford Advanced Surfaces** Based in the UK	It designs, develops and manufactures advanced materials for surface modification, adhesion promotion and nanomaterial applications based on its highly reactive chemical core platform technology called "Onto". They are using the Onto technology platform to chemically functionalize graphene
20	**QuantumWise**	Provides software solutions for development of nanotechnology. The company is working in close collaboration with the Nano-Science Center at the Niels Bohr Institute of Copenhagen University
21	**Samsung Group** based in South Korea	It is a multinational conglomerate company involved with electronics, mobile phones, displays (LCD, Plasma and OLEDs), materials, insurance, finance, advertising, heavy industry and more.
22	**Sony** based in Japan	Sony is one of the leading consumer electronics companies in the world. Sony is developing graphene technologies—including a roll-to-roll production process that can produce graphene sheets up to 100 meters in length
23	**Strategic Energy Resources** is an Australian based company	The company is exploring and developing land with prospective large discoveries in Western Australia, South Australia and Victoria. It is also involved with graphene
24	**Valence Industries** an Australian industrial manufacturing company	It produces high grade flake graphite products

9.2.3 End-User Markets and Target Customers

The Market and target customers depend on graphene-based products such as transparent conductive thin film, displays, composites, coatings, batteries, biomedicine, supercapacitors, adhesives, filtration, catalysts, textiles, sensors, etc. Some of the companies that might be on the verge of starting product based graphene are enumerated here.

9.2.3.1 The Automotive Industries

Automotive industries require catalysts, temperature control materials, self-cleaning coatings and wear-resistant coatings. For this, the target customers are Dow Chemicals and DuPont.

There are efforts to produce cheaper and stronger graphene paper. The graphene paper is being synthesized by heat treatment and has very high bending rigidity and hard mechanical properties. As compared to steel, the prepared graphene paper is six times lighter, with five to six times lower density, two times harder with 10 times higher tensile strength, and 13 times higher bending rigidity.This graphene paper may find its way into aviation industries also. Being lighter, it will use less fuel.

9.2.3.2 Electronic Industries

Industries like Samsung, IBM, Nokia, etc., will be the target customers for transparent conductors, thermal management, displays and printed electronics.

9.2.3.3 Aerospace Industries

Aerospace industries have shown much interest in getting products that can be used in structural monitoring material, flame-retardant materials, conductive and anti-icing coatings, etc. Here also, DuPont as well as BASF are good target customers.

The present day target of all aerospace industries are to reduce the weight, thus making planes that save fuel consumption. Moreover, apart from being tough and crack resistant, there is a need for material that can withstand large temperature ranges (–50° to + 50°C). For this purpose, Polygraph has already scaled up the production of graphene-reinforced thermosetting polymers for composite, coating and adhesive applications. The graphene-reinforcement thermosetting polymers meet these criteria. Use of epoxy adhesives would be useful in aero engine parts, which demand higher thermal stability. Such advanced coating is expected to reduce the costs. To reach these goals, Polygraph has a consortium of partners that include:

- Net composites Ltd to develop new composite materials
- Imerys graphite and carbon to supply novel graphite grades
- Avanzare innovacion tecnologica S.L. a Spanish company to synthesize graphene
- YTRON Process Technology GmbH and Co. *KG* designs efficient rotor/stator technology, etc.)
- University of Padova, Queen Mary University of London, and Netzsch Feinmahltechnik GmbH are collaborators for fundamental characterization (SEM, TEM, FTIR, DMA TGA, UV-VIS, XRD of products and defining the polymer as well as developing exfoliation and dispersion technology)
- Saireem Sas being a microwave expert company, have also joined hands with Polygraph to develop microwave-assisted production techniques for the production of graphene and also to provide electromagnetic hardware components for microwave and/or microwave plasma processing
- Robnor Resins, specialists in formulation of epoxy and polyurethane resins, will be involved in formulation
- Centro Ricerche FIAT for dissemination and exploitation of activities for auto and aero transport industries

9.2.3.4 Energy Sectors

The energy sector is coming up as a big buyer for graphene-based materials for energy storage, photovoltaics, and filtration. Here, the customer list includes BASF, Bayer, and Dow Chemicals.

Gladiator Consortium, including members from seven European countries, have focused their interest in commercializing graphene-based energy storage materials. This consortium includes eight industries (SURAGUS, SGENIA Soluciones, Organic Electronic Technologies, Horiba Jobin Yvon, Graphenea, Amcor Flexibles, Aixtron and Amanuensis) and six universities and research institutes (Technical University Denmark, National Centre for the Working Environment, Leibniz Institute of Surface Modification, Fraunhofer COMEDD, CEA and Aristotle University of Thessaloniki).

Graphene batteries and electrodes have shown lot of advancement. Northwestern University scientist have already (in 2011) synthesized a graphene electrode that has improved the lithium-ion battery to store 10 times more power, charge 10 times faster and last longer. The anode is made of graphene sheets with many 10–20 nm holes punched in it, and between the two sheets of graphene they have introduced clusters of silicon. The advantage of engineering holes in the graphene sheet is that lithium ion will not have to travel around the outer edge of each graphene layer, rather it can now jump through the holes punched in the graphene. Vorbeck Material is developing lithium-ion batteries with graphene for production. Last year (2013) a Michigan-based company, XG Sciences, Inc., has also launched a Li-ion battery using graphene in anode, which has four times more capacity than conventional anodes. They have been funded by US Department of Energy.

Graphene supercapacitors companies are finding graphene as a promising material for high-performance supercapacitors, because graphene-based supercapacitors show high stability and significantly improved electrical double layer capacitance, and also energy density with fast charging and discharging time at a high current density, due to enhanced ionic electrolyte accessibility in deeper regions. One of the recent advancements is a graphene-based micro-supercapacitor developed by the researchers at UCLA. They have achieved this by inserting silicon clusters between each layer of graphene sheets. Needless to say, these miniaturizations of electronic gadgets are in demand. Moreover, the silicon clusters inserted between each layer of graphene sheets have another advantage. Pure graphene can carry only one Li atom/6C atom, whereas each Si atom can hold 4 Li atoms. Thus, the anode becomes more powerful and can store 30,000 mAh instead of 3,000. It has been shown to possess increased charging speed, i.e., 15 min instead of 2 h. However, so far no ultra-supercapacitor using graphene has come to the market.

There are efforts to produce 3D supercapacitors by scientists at Rice University. In December 2014 they announced that using a LIC process they designed, they have produced 3D supercapacitors. In the LIG process, a computer-controlled laser burns through a polymer to create flexible, patterned sheets of multilayer graphene that may be suitable for electronics or energy storage. They have made supercapacitors with laser-induced graphene on both sides of a polymer sheet. The sections were stacked with solid electrolytes in between to get a multilayer construct with multiple micro-supercapacitors. They are scaling it up for commercial applications. Because they have found that as compared to the lithium-ion batteries, LIG supercapacitors (of the same size) have three times the power performance (speed of energy flow). Moreover, the LIG supercapacitors are flexible and produced in open air at room temperature without the need of a clean room and it is sure to find various flexible electronics applications.

Another joint effort between Lomiko Metals, a Canadian company, and Graphene ESD Corp. is to develop low-cost graphene-based supercapacitor that will be capable of very high discharge currents and can be a versatile energy storage solution for electronics, electric vehicles, and electric grids.

Another use of graphene in the energy sector is in improved wind turbines. The addition of graphene to epoxy composite at a ratio of 0.1% of the weight of the compound increases the strength of composite much more than the material made by using CNT (as suggested by Renselaar Polytechnic Institute). Hence, these composites are now being used in windmill blades for their ultra-light, ultra-strong properties. It is suggested that incorporating graphene within wind turbine technology could be cost-effective for offshore turbines.

9.2.3.5 *Graphene Solar Cells*

Graphene, being highly conductive, completely transparent, very thin and unaffected when applied on thin film of Si, is being considered as the material for solar cells by HZB Institute for Silicon Photovoltaics as presented in their press release:

> To this end, they grew graphene on a thin copper sheet, next transferred it to a glass substrate, and finally coated it with a thin film of silicon. They examined two different versions that are commonly used in conventional silicon thin-film technologies: one sample contained an amorphous silicon layer, in which the silicon atoms are in a disordered state similar to a hardened molten glass; the other sample contained poly-crystalline silicon to help them observe the effects of a standard crystallization process on graphene's properties.
>
> Even though the morphology of the top layer changed completely as a result of being heated to a temperature of several hundred degrees C, the graphene is still detectable.

Recently (2014), a group of scientists at Spain's Universitat Jaume I in Castello and Oxford University, working on photovoltaic and optoelectronic devices, have reported fabricating a graphene-based solar cell with 15.6% efficiency. It is expected to be in mass production soon. This solar cell is made by using a combination of titanium oxide, graphene as a charge collector and perovskite as a sunlight absorber. They have lowered the production cost by manufacturing it at lower temperature (150^0 C) depositing it by a solution-based technique.

DSM Ahead, a Dutch company in collaboration with the technical University Eindhoven, are interested in commercializing a graphene solar cell and are a very close to production.

A team of international researchers at the Department of Energy's SLAC National Accelerator Laboratory, examined the properties of materials that combine graphene with a common type of semiconducting polymer, and found that a thin film of the polymer transported electric charge better when grown on a single layer of graphene than it does when placed on a thin layer of silicon.

The scientists claim that their study is one of the first to measure the charge transport in these materials in the vertical direction—the direction that charge travels in organic photovoltaic devices like solar cells or in light-emitting diodes. A somewhat surprising result of the study was that a polymer film about 50 nanometers thick conducted charge about 50 times better when deposited on graphene than the same film about 10 nanometers thick. The team concluded that by better controlling the thickness and crystalline structure of the semiconducting film, it may be possible to design even more efficient graphene-based organic electronic devices and benefit areas like next-generation photovoltaic devices and flexible electronic devices.

9.2.3.6 *Manufacturing Sectors*

Manufacturing sectors are currently researching various applications and approaches using graphene. They are expected to come up with massive demand for synthesizing graphene as well as graphene-based products such as inks, anti-bacterial, catalysts, barrier coatings, sports goods, and filtration. The companies that are showing interest at the moment are 2M and DuPont.

Manufacturing requires equipment specifically for graphene synthesis and the sector has already entered into producing equipment and machinery for graphene synthesis. 2M Strumenti is marketing Nano CVD systems produced by Moorfield. This equipment is designed to produce conditions that allow for rapid, cost-effective production of graphene and carbon nanotubes (CNT), through the implementation of chemical vapor deposition (CVD) technique. Another 2M product is 2M LAB-CVD, which is a tube-based deposition system. It is configured such that size of graphene can be scaled up from 1" to 6" and 8" large substrate. It has supporting gas panel and digital mass flow controllers. The gas panel can be configured for liquid precursors, with an innovative vaporization system for stable flow. AIXTRON-Nanoinstruments are manufacturing PECVD a plasma-coupling technology-based graphene synthesis equipment called Aixtron-"Black Magic."

Graphene-based conducting ink is finding its use in direct ink jet printing of flexible conductive patterns. It is produced by liquid phase exfoliation of graphite in *N*-methylpyrrolidone. It is used to print thin-film transistors. Vorbeck Materials is making this conducting ink commercially.

Graphene-based 3D printing is a hot topic. Many companies are trying to enter this arena, e.g., Graphene 3D Lab, Graphene Technologies, Stratasys, Grafoid, AGT, Qingdao Unique Products and Kibaran.

Graphene 3D Labs is a spinoff of Graphene Laboratories. It has a joint venture with Lomiko Metals. It has an active R & D unit and pilot plant. It will soon unveil a graphene-based 3D printing material. Filaments infused with graphene can be conductive and much more durable than non-specialized filaments. Graphene-enhanced 3D printing materials can be used in the current generation of 3D printers. They are also trying to develop filaments with magnetic capabilities. According to Dr. Elena Polyakova, founder and COO of Graphene 3D Labs, "In high-end electronics, conductive graphene-infused 3D printing materials may allow manufacturers to print end-products in a one step process. They can print not only the mechanical structure, but also integrated circuit layouts and heat-dissipation structures with the push of a button."

Graphene 3D Labs has recently introduced its 3D printed graphene battery prototype, which is composed of nanoplatelets of graphene that are added to polymers. It can produce the same amount of energy as a common AA battery. These batteries will be integrated into a 3D-printed object. To produce it they have entered into partnership with Stonybrook University and signed a contract with US-based ZeGo Robotics to develop a 3D printer prototype machine using company's conductive graphene filament and other functionally enhanced composite materials.

Qingdao Unique Products is a Chinese company which is already using the material, composed of fiber glass graphene for actual printing. They have exhibited a huge 12x12 meter, 3D printer, during a trade show in China, which they plan to use for printing buildings. The printing material is said to be a fiber glass-graphene composite which should enable it to create very strong objects. They have an ambitious plan to print a replica of Beijing's Temple of Heaven of 7m in diameter and 8 m high.

Graphene Technologies in partnership with Stratasys are also aiming to develop graphene-enhanced 3D-printing materials. Graphene Technologies, a California-based company, have their own patent to synthesize graphene from carbon dioxide in an eco-friendly manner, whereas Stratasys is a leading 3D printing company, worth over $5 billion. The US-Israeli BIRD Foundation will help fund this development effort.

Another industry that has entered into investigating graphene-enabled 3D printing is 3D Graphtech Industries along with Australia's CSIRO.

Another is graphene-based 925 hp sports car by a Spanish car company, Spania GTA. The chassis and body work of this car named GTA Spano is made using graphene, carbon fibre, titanium and kevlar. The company is planning to release only 99 cars at the moment.

Graphene-based anti-bacterial paper and anti-bacterial fabrics are likely to find application in food packaging and other products like masks and bandages, etc. However, it has not yet been commercialized.

Sporting goods companies are interested in producing strong and light equipment and have shown interest in the application of graphene.

Novak Djokovic, the tennis champion, uses a tennis racket made of graphene. This racket is made by HEAD, an Austrian sports equipment making company. HEAD has worked with Taiwan's Industrial Technology Research Institute to develop this racket. The project was funded by Austrian Government.

Zyvex Technologies, a company known for being the world's first molecular nanotechnology company and for developing a CNT and graphene engineered composite material along with ENVE Composites Company, has produced the first bicycle rim comprising nanofiber. This bicycle rim is specially designed for downhill mountain biking. The CNT and graphene composite provides extra toughness and protects from fracture damages.

Haydale a company based in South Wales, UK, is also developing a program to improve the strength and stiffness using graphene-enhanced material so as to produce a strong light weight racing boat. In this venture they have made a collaborative agreement with Alex Thomson Racing sailing team. Haydale's newly acquired subsidiary EPL Composite Solution is working for developing graphene-enhanced material for this purpose.

9.3 Graphene Commercialization

Commercialization of a product and demand for the application of products are interrelated. Unless both co-exist, economics will not meet expectations and commercialization becomes uninteresting for industrialists. Fortunately, with graphene, both prongs are being considered and importance is being given to both equally. It is therefore possible to observe some breakthroughs in graphene science. Nevertheless, there are some teething problems with both factors. Some of them are examined here.

Though there is a need, opportunity and market for graphene and graphene-based products, in the absence of scaled-up mass production of desired-quality graphene (i.e., layers of graphene, band gap or electrical and optical properties), most applications are only slowly coming out of the research labs. Though there are many people working on the commercial applications of graphene, there are still some important graphene inventions that have not yet been successfully commercialized despite demand: Graphene-based holographic optical disc; graphene photo-detectors; DNA-based graphene transistors; and graphene-copper composites.

To meet both these aspects, there is a need for collaborative research towards the development of large-scale graphene production for device integration.

To realize the commercialization of graphene one must know:

a. What are the timelines and roadmaps for these applications? For this supply chain and product development road map needs serious considerations.
b. Will material capability translate into device performance? Verification of graphene property to identify the viable application will be important criteria.
c. How much will the material cost, and to keep the price low, how much would be its demand?
d. In what qualities and quantities will it be available? Tackling quality of the supply to identify necessary assurance steps for commercial production has to be looked into.
e. Is production scalable? Scaling-up and processing will demand assessing the techniques and steps required for mass production and processing scale-up of graphene.
f. Is it economically viable? There is a need to discuss the cost projection after examining the current and projected cost of graphene for comprehensive assessment of investment viability.

9.3.1 Challenges in Graphene Commercialization

One of the challenges is the variation in properties (e.g., crystallite size, sample size, charge carrier mobility, reproducibility in number of layers of graphene formed by each batch, etc.) of graphene fabricated by different methods.

9.3.1.1 Producing Desired Band Gap

The biggest challenge in commercialization of graphene is in producing graphene of desired band gap. As we know, graphene is highly conductive or rather too conductive for many electronics applications and it has no band gap. It can conduct electrons or holes at a speed that is 100 times faster than any other materials, which is nearly the speed of light. To overcome the band gap problem, artificial band gap is being introduced into graphene by doping (i.e., by introducing or converting some of sp^2 converting into sp^3 carbon) into its surface with chemicals or by making nanoscale ribbons. But these complex methods make it difficult to adapt for mass production because the process becomes difficult to get a reproducible product.

Another challenge is in fabricating defect-free sheets of graphene at a commercial level. Production of large quantity of pure graphene (i.e., defect-free) is turning out to be very difficult. Many industries are focussing on producing small area flakes of graphene at large scale to reduce the incidence of defects.

Another important challenge is deposition of graphene on an appropriate substrate so that it could be for the specific application.

9.3.1.2 High Production Cost

High production of graphene is also a major obstacle in the commercialization of graphene. The current cost of 50x50 cm mono-layer graphene thin film on Cu-foil is $263 and on PET is $819. At the same time, 5–8 nm thick graphene nanoplates are being sold by XG Sciences at $219 to 229/kg.

According to Lux Research (Table 9.1), the cost of graphene was $ > 250/kg in 2011 and will go down to $10–15/kg by 2015. If graphene is to replace silicon, it has to be not only very cost-effective but has to offer extraordinary performance, because a billion dollars of investments has been made in manufacturing infrastructure of silicon around the globe and company is likely to replace it with graphene, unless it is profitable.

However, one of the graphene-based applications, i.e., water desalination plants, has been envisaged to be 15–20% cheaper to run than traditional ones, because their lower pressure means they need less energy and they have much greater throughput. Nevertheless, there are some basic problems associated with this system in the design of the graphene monolayer membrane and its durability.

9.4 Economics of Graphene and Graphene-related Products

The present day belief of industries and government is that graphene has the power to change economics. Fueled with this belief, the European

Commission in 2013 announced graphene as a billion euro technology flagship, signaling a continent-wide effort to advance the development of the revolutionary material to mass-market applications. With this financial support for research and development, graphene-based technologies and associated business opportunities have never been so promising. However, the question is: Will graphene deliver the goods as expected? At the moment, the cost of graphene is very high. An online shop, Graphene Supermarket, offers a kilogram of pristine graphene at $40,000. However, it must be mentioned here that the average graphene nanoplate selling price that was approximately US$250/kg in 2011 has now gone down to US$20/kg.

So far as market revenue estimate is concerned, varied predictions have been made by different agencies (Table 9.6).

According to the vision of Kitty Cha of BASF, graphene commercialization will start with composite materials and electronic inks. For such applications, graphene can be extracted from graphite flakes, which greatly simplifies its large-scale production. A word of caution is needed. Where will the graphite be obtained? Is it globally available or does it depend upon some specific region of the world? Once commercialization is thought of from graphite, geographical location of its availability becomes an important aspect in controlling the price of graphene. Many scientists believe that technologies based on large graphene films such as displays, photovoltaic cells, batteries or supercapacitors will emerge. Amaia Zurutuza, scientific director of Graphenea, a premier company involved with graphene, has illustrated the potential of graphene for the energy industry. Their company is researching batteries and light harvesting.

So we see that the opinions differ when it comes to the long-term applications. Graphene-based transistors are mentioned often, but according to Tomas Palacios, associate professor of electrical engineering at MIT, it will be difficult to compete against the silicon industry. He said, "the economics are against us, the work of thousands of intelligent silicon engineers is against us, and the physics is against us." Similarly, S. Park from Samsung Electronics feels that the thin ITO films are as flexible as graphene, thus contradicting the common idea that only graphene is capable of making flexible displays.

It must be said here that people have started thinking beyond graphene as well. In fact, graphene might just be the tip of the iceberg. The "good idea" that graphene developers are hunting for might include, not only graphene, but a whole class of two-dimensional crystals like boron nitride and transition metal dichalcogenides. All graphene developers agree: If graphene does not replace current technologies, it will lead to novel technological concepts and devices.

Table 9.6 Projected global graphene market value by different agencies

Projected by	**In the year**	**Graphene Market**
"Graphene: Technologies, Applications and Markets" released by BCC	2015	$67 million
	2020	$675.1 million
Graphene-Markets: http://www.idtechex.com/research/reports/Graphene-markets-technologies-and-opportunities-2014-2024-000390.asp (*Source: IDTechEx*)	2014	$ 20 million
	2024	$ 390 million
Allied Market Research	2020	$149.1 Million
Projected by	**In the year**	**Production Volume**
World Market for Graphene by the Future Markets Inc.	2010	28 tones
	2017	573 tones
Projected by	**In the year**	**Production Cost**
Lux Research	2011	$ > 250/kg
	2015	$ 10–15/kg
Data from	**In the year**	**Patents Filed**
Thomson Innovation database https://www.thomsoninnovation.com	2008	5
	2011	141

The commercial/economic viability of graphene-related industries can be assessed from the fact that a new non-profit organization, Graphene Stakeholders Association (GSA), was formed in April 2013 to promote responsible development and commercialization of graphene and graphene-enabled products. The GSA is headquartered in Buffalo, NY; it is founded by the joint efforts of Keith Blakely, of the advanced materials and nanotechnology community, Dr. Alan Rae, CEO of the Nano Materials Innovation Center, Vincent Caprio, Executive Director of the NanoBusiness Commercialization Association and Stephen Waite, Managing Partner of SoundView Technology Group. Mr. Waite and Mr. Blakely are the co- Executive Directors of the GSA.

Focus Metals, a Canadian mining company, and Grafoid, a Canadian materials company, have the process to produce cheap graphene using unprocessed graphite ore as a raw material. However, the graphene economics is unclear at the moment because most of the graphene production is patented or is a trade secret. There are other mining companies also looking for potential intellectual property advancements to produce graphene using their graphite ore, e.g., Northern Graphite Corp. who are supplying extra large flake graphite to Grafen Chemical Industries for graphene research.

One of the great concerns graphene companies have is regular availability of raw material. More than 70% of the mineral is produced in China, and the graphite producers of China are forcing the government to declare same protection to graphite as given to rare earth elements. If this happens, then one can visualize a disruption in graphite supply causing increased prices. However, despite these hurdles, companies and governments are pumping billions of dollars into graphene research and its commercialization.

9.5 Graphene and the Future Possibilities

Science and fantasy have always been hand-in-glove. Jules Gabriel Verne fantasized a trip to the moon, but he was not the only one. Edgar Rice Burroughs, E. R. Eddison, John Carter of Mars series, David Lindsay's *A Voyage to Arcturus*, are also examples of science-based fantasy. Fantasy is the realization of far-fetched concepts. Journey from fantasy to reality could be long and laborious; but no doubt it has always been rewarding, like flying in the sky.

Who imagined that this miraculous material graphene, which was first isolated from the tip of a pencil, will be the material of the future. Is its transparent, lightweight, flexible, strong and very thin (one atom thick) and many other unique properties, such as exceptional charge transport, thermal, optical, and mechanical properties, that has evoked the researchers from all the branches of science, instigating the industrialists across the globe to start imagining to have possibility of building future technologies based on it? In one word, "Yes" to all of it. It is believed that it has the possibility to replace the silicon and ITO in the very near future because it has all the positive characteristics of exotic alloys of metals and plastics that have in the past revolutionarized modern-day living.

With applications envisaged for foldable touch screens for mobile phones; as drug-delivery vehicles, protective coating, food packaging, wind turbines, faster computer chips and broadband; batteries of very high capacity and see-through high efficiency solar panels built directly into a colorful screen; the technological revolution using graphene is only a step away and is on the verge of becoming a reality.

Moreover, unlike other materials in which one absorbed photon generates one electron, the one absorbed photon by graphene produces many excited electrons and generates larger electrical signals (Frank Koppens). Graphene can thus capture generated electricity. This property of graphene has opened the door for many near future applications such as in Li-ion battery to make a longer lasting battery. Plus the perfect crystallinity as well as high conductivity of graphene promises to have great applicability in electronics and electronic devices such as diodes and transistors.

Specialties of graphene are highlighted in chapters 1 and 2, so therefore to avoid repetition, only some special applications are discussed here. Though graphene possesses many special properties, four basic properties of graphene seem to capture industrialists the most. These are its almost 90% transparency up to 500nm, large surface area (2630 m^2 g^{-1}), thermal conductivity ($\approx$5000Wm K^{-1}) and its very low electrical resistivity (10–30 Ω/square) for both electrons as well as for holes; whereas graphite shows 104 Ω^{-1} cm^{-1}electrical conductivity, 3000 W/mK thermal conductivity and 200,000 cm^2 $V^{-1}s^{-1}$ electron mobility. These properties of graphene have made it superior to graphite as well as silicon that has 149 $W{\cdot}m^{-1}{\cdot}K^{-1}$ thermal conductivity and 10Ω/m^3 electrical resistivity at 20°C.

According to scientists at Rice University in collaboration with Moscow State University, another great possibility of application of graphene is cleaning up the radioactive waste like the polluted water of Fukushima. They have shown that graphene oxide flakes that are soluble in water can quickly clump together radionuclides waste present in contaminated water and condense them into solids, thus making the disposal easier.

9.5.1 Flexible Electronic Screens

One of the first future possibilities to become reality is likely to be the flexible electronic screens that will be used for mobile phones, for example.

Graphene, with its electrochemically stable property combined with its flexibility and transparency makes it most suitable material for making light-emitting electrochemical cells (LECs). There have been some success at bench level and it may soon become a marketable product. The electronic giant Samsung is one of many companies interested in this product. Apple, which has produced the first touchscreen iPhone using ITO, was expected to jump into this technology, but they seem to be lagging behind Samsung, who have 38 patents related to graphene as compared to only two by Apple.

BASF and the National University of Singapore have joined hands to develop graphene for organic electronics devices, such as OLED devices and OPVs.

9.5.2 Graphene Composites of Very High Mechanical Strength

Another likely near future application could be making graphene-incorporated composite material of very high mechanical strength that can be used in buildings, cars, ships, boats, aircraft, spacecrafts, robots, etc. No wonder Michiko Kaku said that the graphene will change the way we live.

It will be hard not to mention the tennis rackets made by using graphene that are used by champion Maria Sharapova and the world's number one tennis player Novak Djokovic. It is a strong and very light racket, but due its proprietary nature, many details are not available. But, there may soon be cricket bats or hockey sticks using graphene.

9.5.3 Graphene to Replace Flash Memory of SD Cards

It is also hoped that graphene will replace the flash memory of SD cards and the metal antennae that picks up radio signals.

9.5.4 Next Generation Speakers

Another future possibility seems to be replacement of conventional speakers that drive sound vibrations by pumping the air. Graphene has been found to achieve the same by using heat. Graphene being very thin uses almost no current to get heated and it cools down instantly. So through a graphene sheet, audio-frequency currents can be easily pumped to generate sound waves and high fidelity audio speakers can be made.

Ji Won Suk at the University of Texas is developing a "Thermophone" with a transparent speaker over the whole screen.

9.5.5 Faster Computer Chips and Broadband

Industries involved in manufacture of electronic components are continually coming up with products having ever-shrinking size. Efforts are on to squeeze in more components on a chip < 1 cm across. The crystalline perfection of graphene combined with its higher conductivity than silicon are now showing promise to develop electronic devices such as diodes and transistors at the molecular level.

IBM has already proclaimed the fabrication of a wafer-scale graphene circuit. This circuit is a broadband radio-frequency mixer (an essential component of TVs, phones and radio) by exploiting the high speed of electrons in graphene.

This property of graphene is expected to be exploited for its use in the high frequencies of mobile phone transmissions—both in the receiver and in generating the signal for the base station. IBM has already made graphene components that operate up to 150 GHz, well above the frequencies currently used by mobile phones; this will open up new portions of spectrum for cell phone coverage and improved connectivity.

The Pitfalls: There are chances of things going wrong at every step; however, we must not forget that the first attempt of making electronic components using silicon was a total flop, but success was achieved in the end. As compared to silicon, the progress with graphene is rather rapid.

The importance of graphene can be seen by the concerns of many environmentalists who have already started to study the future impact of graphene and graphene-based materials. Researchers at the University of California are concerned in knowing what would happen if graphene is released into the environment. Will it be a future pollutant? This is because graphene oxide nanoparticles quickly spread on the surface of water, and if it finds its way into a lake, river or sea, it may harm plants, animals and perhaps even humans. Whenever a new branch of science has mushroomed there have been environmental concerns. Accordingly, attention has been paid to research and regulations. Such activities are likely to be taken up for graphene as well.

9.5.6 Super-Strong Bulletproof Body Armor Using Graphene

Researchers from the University of Massachusetts have shown that when a bullet impacts single-layer graphene sheets, it dissipates kinetic energy by stretching into a cone shape at the bullet's impact point, then radially cracks outward. They suggested that although the cracks formed, the graphene body armor still performed twice as well as Kevlar and endured 10 times the kinetic energy that steel can. They are trying to abate the cracking problem by using multiple layers of graphene or by making a composite structure. Graphene, being very light and having high mechanical strength, is obviously potentially fit for making body armor.

9.5.7 Graphene Drones

DJI, a Chinese-based international company, is launching a graphene drone which will be a multi-rotor aircraft that uses graphene in the frame, arms and props. The drone will be lightweight with strong stiffness, and the battery will get higher capacity, faster loading, and longevity. However, since graphene batteries have not yet been commercially used, this may cause a delay in introducing graphene drones.

9.6 Graphene and Fantasies

Graphene is seen as the miracle material of future by the people belonging to all fields of science and engineering. Their enthusiasm with this material has even impelled them to fantasize the various applications of graphene. Even the judges of the Nobel Prize have remarked that most of graphene's practical applications "exist only in our fantasies, but many are already being tested."

The fantasy and imagination of Rebecca Clancy that "you could wheel your trolley between two sensors that would scan every product immediately and give you your final bill in the blink of an eye" (Daily Telegraph, London, August 18, 2013) would surely be a fantasy of housewives waiting for it to become a reality.

Many recent discoveries of properties of graphene are opening doors for scientists to fantasize about a future with graphene encompassing all the fields; such as:

- A study on the dynamics of graphene electrons in a magnetic field has yielded extraordinary results that make scientists at the Helmholtz-Zentrum Dresden-Rossendor, Germany, believe could enable a new generation unique laser technology.
- A graphene-infused nylon material for use in 3D printing is being envisaged by graphene 3D Lab, a US based company.
- A European Union–funded project is focusing on the development of a graphene-based scanner, which can be used to reveal hidden aspects of artwork and other historic objects on canvases, like preliminary sketches or former paintings, as well as the way brushstrokes were done. This information can shed light on how great art was made, and also be helpful in restoration of paintings.
- Researchers from Stockholm University have developed a super-insulator and fire retardant foam; by freezing a mixture of graphene oxide, cellulose nanofibers and clay nanorods. It can be used the in future for house insulation.
- An optical sensors has already been patented by R & D unit of Nokia. It is a graphene-based photodetector. The company is

now wishing to develop graphene devices to create the basis for a range of optical sensor products.

- The day is not far off when one can plug his phone and within five seconds the phone will be charged.
- Similar graphene batteries, which are nontoxic, inexpensively produced and charge superfast, can be used in electric cars and will charge as fast as filling a car with gas.

There is a deluge of research news regarding the unique property-based applications of graphene, coming every month to the market.

It has been observed that whenever a new product is brought into focus light, scientists start researching with the material intensely with a hope that the new material will give the answer to many existing problems of science. This has been seen with the development of fullerene. Research was carried out with a hope that it can break many deadlocks of science. After few years of research, scientists realized that fullerene does not have any specific application other than academic interest. Then came the discovery of carbon nanotubes, carbon nanofibers and many other forms of nanocarbon. Scientists worked very diligently on these materials and fortunately many problems have been resolved, providing, the cost of its preparation is lowered. In fact, trends suggest that nanocarbon can be synthesized cheaply and can be useful materials for many industrial applications.

Similarly, with the discovery of graphene and graphane, scientists have delved into work on this material very methodically and robustly. Many applications at the lab scale have been shown. Efforts are also being made to produce large quantities of large-size graphene. The cost of production has also been lowered. Scientists have many hopes with this material and it is believed that graphene may be able to remove many deadlocks of science. Only time will be able to really tell how much graphene will be useful. Will industrialists forget graphene like fullerene or will they develop graphene

(a) (b) (c)

Fig. 9.1 (a) Ceramic from the third century (b) Setup for 3D acquisition with a structured light scanner (c) 3D model resulting from raw data before post-processing. (*Taken from Graphene-Info.com newsletter December 2014*)

to catch up with carbon nanotubes (or surpass carbon nanotubes)? Only time will tell.

9.7 Summary

In this chapter, we have examined the challenges the industry and scientists face to make graphene a material for the twenty-first century. A list of companies working in the various aspects of graphene has also been given and discussed. It has been the object of this chapter not only to show the golden positive aspects of graphene but also to alert scientific organizations and researchers that only time will be able to tell in which areas of science graphene will flourish the most. It would be a real miracle if graphene could find its way in all areas where scientists have dwelt to carry research.

References

1. Abhilash M.; *International J. of Pharma and Bio Sciences*, Vol. V1(1) p. 1, 2010
2. Ahn S., Cha J., Myung H., Kim S-M. and Kang S.; *Appl. Phys. Lett.*, Vol. 89, p. 213101, 2006
3. Ahn S. H. and Guo L.; *J. Adv. Mater.*, Vol. 20, p. 2044, 2008
4. Allen M. J., Tung V. C. and Kaner R. B.; *Chem. Rev.*, Vol. 110, p. 132, 2010
5. Anderson M.S.; *Appl. Phys. Lett.*, Vol. 76, p. 3130, 2000
6. Bae S., Kim H., Lee Y., Xue X., Park J-S, Zheng Y, Jayakumar B., Im D., Lei T., Song Y., Kim Y.J., Kim K.S., Özyilmaz B., Ahn J-H., Hong B.H. and Ijima S.; *Nat. Nanotechnol*, Vol. 5; p. 574, 2010
7. Bagri A., Kim S.P., Ruoff R.S. and Shenoy V.B.; *Nano Lett.*, Vol. 11, p. 3917, 2011
8. Bai J. W., Zhong X., Jiang S., Huang Y. and Duan X. F.; *Nat. Nanotechnol.*, Vol. 5, p. 190, 2010
9. Baker S. N and. Baker G. A.; *Angew. Chem., Int. Ed.*,Vol. 49 p. 6726, 2010
10. Balandin A.A., Ghosh S., Bao W., Calizo I., Teweldebrhan D., Miao F. and Lau N., *Nano Lett.*, Vol. 8, p. 902, 2008
11. Balapanuru, J., Yang, J.X., Xiao, S., Bao, Q.L., Jahan, M., Polavarapu, L., Wei, J., Xu, Q.H. and Loh, K.P.; *Angew. Chem. Int. Ed.*, Vol. 49, No. 37, p. 6549, 2010
12. Beck M., Persson F., Carlberg P., Graczyk M., Maximov I., Ling T.G.I. and Montelius L.; *Microelectron Eng.*, Vol. 873, p.73, 2004
13. Bera D., Lei Qian, Teng-Kuan Tseng and Holloway P. H.; *Materials*, Vol. 3, No. 4, p. 2260, 2010
14. Berger C., Song Z., Li T., Li X., Asmerom Y. O., Feng R., Dai Z., Marchenkov A.N., Conrad E.H., First P.N. and de Heer W.A.; *J. Phys. Chem. B*, Vol. 108, No. 53, p. 19912, 2004
15. Berger C., Song X., Li X., Wu X., Brown N, Naud C., Mayou D., Li T., Hass J., Marchenkov A.N., Conrad E.H., First P.N and de Heer W.A.; *Science*, Vol. 312, No. 5777, p. 1191, 2006
16. Bernal, J. D., *Proceedings of the Royal Society of London*, Vol. 113, No. 763, p. 117, 1928
17. Bernardi M., Palummo M. and Grossman J.C.; *Nano Lett.*, Vol. 13, No. 8, p. 3664, 2013
18. Blake P., Brimicombe P.D., Nair R.R., Booth T.J., Da Jiang, Schedin F., Ponomarenko L.A., Morozov S.V., Gleeson H.F., Hill E.W., Geim A.K. and Novoselov K.S.; *Nano Lett.* Vol. 8, p. 1704, 2008
19. Boehm H. P., Clauss A., Fischer G. and Hofmann U.; *Proceedings of the Fifth Conference on Carbon.* Pergamon Press, 1962
20. Boehm H. P, Clauss A., Fischer G. O. and Hofmann U.; *Z. Naturf.*, Vol. 17, p, 150, 1962

21. Bolotin K.; Ghahari, Fereshte J. E., Shulman, M.D., Stormer H. L. and Kim P.; *Nature*, Vol. 462, No. 7270, p.196, 2009
22. Bommel A. J. Van, Crombeen and A. Van Tooren; *Surface Science*, Vol. 48, No. 2, p. 463, 1975
23. Bonaccorso F., Sun Z., Hasan T. and Ferrari A. C.; *Nature Photon*, Vol. 4, p. 611, 2010
24. Brodie B. C.; *Philosophical Transactions of the Royal Society of London*, Vol. 149, p. 249. 1859.
25. Bouchard E., Lavenac J, Roux J.C., Langlais F. and Delhaès P.; *Chemical Vapor Deposition*, Vol. 7, Issue 3, p. 125, 2001
26. Boukhvalov D. W and Katsnelson M. I., *Appl. Phys. Lett.*, Vol. 95, No. 2. 2009
27. Bourlinos A.B., Stassinopoulos A., Anglos D., Zboril R., Karakassides M.I. and Giannelis E.P., *Small*, Vol 4, p. 455. 2008
28. Britnell L., Gorbachev R.V., Jalil R., Branson D. B., Schedin F., Katsnelson M.I., Eaves L., Morozov S.V., Mayorov A.S., Peres N. M. R., Castro Neto A.H., Leist J., Geim A.K., Ponomarenko L.A. and Novoselov K.S., *Nano Letters*, Vol. 12, No. 3, p. 1707, 2012
29. Brumfiel G., *Nature*. 2009. doi:10.1038/news.2009.367
30. Bullis K., *Technology Review*, (Cambridge: MIT Technology Review, Inc). Retrieved 2008-01-28, 2008
31. Bunch J. S., Verbridge S. S., Alden J. S., van der Zande Arend M, Parpia J.M., Craighead H. G. and McEuen P.L; *Nano Lett.*, Vol. 8, Vol. 8, p. 2458, 2008
32. Cai W., Zhu Y., Li X., Piner R.D. and Ruoff R.S.; *Appl. Phys. Lett.*, Vol. 95, p. 123115, 2009
33. Cai W., A.L. Moore, Y. Zhu, X. Li, S. Chen, L. Shi, R.S. Ruoff; *Nano Lett*. Vol. 10. p. 16451, 2010
34. Cai J., Ruffieux P, Jaafar R., Bieri M., Braun T., Blankenburg S., Muoth M, Seitsonen A.P., Saleh M., Feng X., Müllen K. and Fasel R.; *Nature*, Vol. 466, p. 470, 2010
35. Castro Neto A. H., Guinea F., Peres N. M. R., Novoselov K. S. and Geim A. K.; *Rev. Mod. Phys.*, Vol. 81, p. 109, 2009
36. Chang Y. K. and Hong F. C.; *Nanotechnology*, Vol. 20, p. 195302, 2009
37. Chang H.X., Tang L.H., Wang Y., Jiang J.H. and Li J.H.; *Anal. Chem.*, Vol. 82, p. 2341, 2010
38. Charlier J.-C., Eklund P.C., Zhu J., Ferrari A.C., *Electron and Phonon Properties of Graphene: Their Relationship with Carbon Nanotubes*, in a edited book: A. Jorio, G. Dresselhaus, M. S. Dresselhaus (Eds.): Carbon Nanotubes, Topics Appl. Physics 111, 673–709 (2008), Springer-Verlag Berlin Heidelberg 2008
39. Chen H.L., Chuang S.Y., Cheng H.C., Lin C.H. and Chu T.C.; *Microelectron. Eng.*, Vol. 83, p. 893, 2006
40. Chen Z., Wencai R., Liu B, Gao L., Pei S., Wu Zhong-Shuai, ZhaoJ. and Cheng Hui-Ming.; *Carbon*, Vol. 48, p. 3543, 2010

41. Chen S., BrownL., Levendorf M., Weiwei C., Sang-Yong Ju; Edgeworth J., Li X., Magnuson C.W., Velamakanni A., Piner R.D., Kang J., Park J. and Ruoff R.S.; *ACS Nano,* Vol. 5, No.2, p. 1321, 2011
42. Chen Y.Y., Li S.J., Huang I.J., Pan D.C, *Inorganic Chemistry*, Vol. 52, p. 7819, 2013
43. Choi C.H., Chung M.W., Kwon H.C., Park S.H. and Woo S.I., *J. Mater. Chem. A,* Vol. 1, p. 3694, 2013
44. Choi Y., Ahn J. and Park J.H.; *J. of Emerging Technology and Adv. Engineering,* Vol. 3, No. 12, 2013
45. Chou S.Y., Krauss P.R. and Renstrom P.J.; *Appl Phys Lett*, Vol. 67, p. 3114, 1995
46. Chou S.Y., Krauss P.R. and Renstrom P.J.; *J Vac Sci Tech B,* Vol. 14, p. 4129, 1996
47. Chou S.Y., Keimel C. and Gu J.; *Nature,* Vol. 417, p. 835, 2002
48. Chuan Xiu-Yun; *Int. Nano Letters*, Vol. 3, p. 6, 2013
49. Cohen-Tanugi D. and Grossman J.C.; *Nano Lett.*, Vol. 12, No. 7, p. 3602, 2012
50. Cote L. J., Rodolfo Cruz-Silva and Huang J.; *J. Am. Chem. Soc.*, Vol. 131, p. 11027, 2009
51. Crassee I., Levallois J., Walter A.L., Ostler M., Bostwick A. and Rotenberg E.; *Nature Physics,* Vol. 7, p. 48, 2011 , doi:10.1038/nphys1816
52. Das A., Chakraborty B. and Sood A.K.; *Bull.Mater. Sci.*, Vol. 31, No. 3, p. 579, 2008
53. Dato A., Radmilovic V., Lee Z., Phillips J. and Frenklach M.; *Nano Lett.,* Vol. 8, No. 7, p. 2012, 2008
54. Dean C.R., Young A.F., Meric I., Lee C., Wang L., Sorgenfrei S., Watanabe K., Taniguchi T., Kim P., Shepard K. L. and Hone J.; *Nat. Nanotechnol.* Vol. 7, No. 9, p. 693, 2010
55. de Heer Walt A., Berger C., Wu X., First P.N., Conrad E.H., Li X., Li T., Sprinkle M., Hass J., Sadowski M.L. and Martinez P.G.; *Solid State Communications,* Vol. 143, No. 1–2, p. 92, 2007
56. DiVincenzo D.P. and Mele E.J., *Phys Rev. B,* Vol. 29, p. 1685, 1984
57. Demchenko A.P.; *Cytotechnology,* Vol. 65, No. 2, p. 157, 2013
58. Dong Y.Q, C.Q. Chen, X.T. Zheng, L.L. Gao, Z.M. Cui, H.B. Yang, C.X. Guo, Y.W. Chi and C.M. Li.; *J. Mater. Chem.*, Vol. 22: p. 8764, 2012
59. Y. Dong, H. Pang, H. B. Yang, C. Guo, J. Shao, Y. Chi, C. M. Li and T. Yu; *Angew. Chem., Int. Edn.,* Vol. 52, p. 7800, 2013
60. Dongxing Y., Velamakannia A., Bozoklub G., Parka S., Stollera M., Pinera R.D., Stankovichc S., Inhwa J., Fieldd D.A., Carl A. Ventrice Jr.d and Ruoffa R.S.; *Carbon,* Vol. 47, p. 145, 2009
61. Dresselhaus M. S. and Dresselhaus G.; *Adv.Phys.* ; Vol. 51, p. 1, 2002
62. Dresselhaus M.S., Dresselhaus G., Jorio A., Souza Filho A.G. and Saito R., *Carbon,* Vol. 40, p. 2043, 2002
63. Dubin S., Scott G. and. Kaner R.B.; *ACS Nano*, Vol. 4, No. 7, p. 3845, 2010

64. Duhee Y. Hakseong K., Sang Wook Lee and Hyeonsik C.; *Phys. Rev. B*, Vol. 83, No. 8, p. 081419(R), 2011
65. Eda G., Lin Y. Y., Mattevi C., Yamaguchi H., Chen H.A., Chen I.S., Chen C.W. and Chhowalla M.; *Advanced Materials*, Vol. 22, No. 4, p. 505, 2010
66. Elías A. L., Botello-Méndez A. R., Menese-Rodríguez D., González V. J., Ramírez-González D., Ci L., Muñoz-Sandoval E., Ajayan P. M., Terrones H. and Terrones M.; N*ano Lett.*, Vol. 10, p. 366, 2010
67. Enoki T., Suzuki M. and Endo M.; *Graphite Intercalation Compounds and Applications,* Oxford University Press, U.K., 2003
68. Faye D., Lefevre J.P., Delaire J.A. and Leray I.; *J. Photochem. Photobiol. A-Chem.*, Vol. 234, p. 115, 2012
69. Feng L. and Zhang W.X.; *AIP Advances,* Vol. 2, p.042138, 2012; http://dx.doi.org/10.1063/1.4766937
70. Feng J., Li W., Qian X., Qi J., Qi L. and Li J.; *Nanoscale,* Vol. 4, p. 4883, 2012
71. Feng L., Wu L., Wang J., Ren J., Miyoshi D., Sugimoto N. and Qu X.; *Adv. Mater.,* Vol. 24, p. 125, 2012.
72. Fernandez-Merino M.J., Guardia L., Paredes J.I., Villar-Rodil S., Solis-Fernandez P. and Martinez-Alonso A.; *J Phys Chem C*, Vol. 114, No. 14, p. 6426, 2010
73. Feynman Richard. "Plenty of Room at the Bottom"—Historical lecture given at the *American Physical Society in Pasadena* on December 1959
74. Fletcher A.J.; Carbon, http://personal.strath.ac.uk/ashleigh.fletcher/carbon.htm
75. Frost R., Jönsson G.E., Chakarov D., Svedhem S. and Kasemo B.; *Nano Lett.*, Vol. 12, No. 7, p. 3356, 2012
76. Fujita M., Wakabayashi K., Nakada K. and Kusakabe K.; *Journal of the Physics Society Japan,* Vol. 65, No. 7, 1996
77. Gao W., Alemany L.B., Ci L. and Ajayan P.M., *Nat. Chem.;* Vol. 1, No. 5, p. 403, 2009
78. Geim A. K. and. Novoselov K. S; *Nat. Mater., Vol.* 6, p. 183, 2007
79. Geim A. K. and MacDonald A. H., *Phys. Today*, Vol. 60, p. 35, 2007
80. Geim A.K. and Novoselov K.S.; *Nature Materials*, Vol. 6, No.3, p. 183, 2007
81. Geim A. K.; *Science*, Vol. 324, p. 1530, 2009
82. Goki E., Giovanni F. and Chhowalla M.; *Nature Nanotechnology*, Vol. 3, p. 270, 2008
83. Goki E., Lin Y.Y., Mattevi C., Yamaguchi H. and Hsin-An Chen; *Adv. Mater.*, Vol. 22, p. 505, 2010
84. Gorbachev R. V., Riaz I., Nair R. R., Jalil R., Britnell L., Belle B. D., Hill E. W., Novoselov K. S., Watanabe K., Taniguchi T., Geim A. K. and Blake P.; *Small,* Vol. 7, No. 4, p. 465, 2011
85. Graeme W, Seger B. and Kamat P.V.; *ACS Nano*, Vol. 2, No.7, p. 1487, 2008
86. Graeme W, Seger B. and Kamat P.V.; *Physical Chemistry B,* Vol. 110, p. 8535, 2006

87. Grigaliūnas V., Tamulevičius S., Muehlberger M., Jucius D., Guobienė A., Kopustinskas V. and Gudonytė A.; *Science*, Vol. 253, p. 646, 2006
88. Guang-Xin Ni, Zheng Y., Bae S.,Chin YawTan, Kahya O., Wu J., Byung Hee Hong, Kui Yao and Barbaros Ozyilmas; *ACS Nano*, Vol. 6, No. 5, p. 3935, 2012
89. Guo L.J. and Ahn S.H; US Patent 2011, 8:027,086.
90. Habiba K., Makarov V. I., Avalos J., Guinel M.J.F, Weiner B.R. and Morell G.; *Carbon*, Vol. 64. p. 341, 2013
91. Hackley J., Ali D., DiPasquale J., Demaree J. D. and Richardson C. J. K.; *Appld Phys Lett,;* Vol. 95, No. 13, p.133114, 2009
92. Haatainen T., Majander P., Riekkinen T. and Ahopelto J.; *Microelectron Eng.*, Vol. 83, p. 948, 2006
93. Haatainen T., Majander P., Mäkelä T., Ahopelto J. and Kawaguchi Y.; *Jpn J Appl Phys*, Vol. 47, p. 5164, 2008
94. Han Melinda Y., Barbaros Ozyilmas, Zhang Y. and Kim P.; *Phys. Rev. Lett.*, Vol. 98, p. 206805, 2007
95. Hass J., de Heer W.A. and Conrad E.H.; *J. Phys. Cond. Matter*, Vol. 20, p. 323202, 2008
96. Hayazawa N., Inouye Y., Sekkat Z. and Kawata S.; *Opt. Commun.*, Vol 183, p.333, 2000
97. He H., Klinowski J., Forster M. and Lerf A.; *Chem. Phys. Lett.*, Vol. 287, p. 53, 1998
98. He S.J., Song B. and Li D.; *Adv. Funct. Mater.*, Vol. 20, p. 453, 2010
99. Hernandez Y., Nicolosi V., Muatafa L., Bligheee F.M., Sun Z., De S. McGovern I.T., Holland B., Byrenes M., Gun'ko Y., Boland J. J., Peter N., Duesfberg G., Krishnamurthy S., Goodhue R., Hutchinson J., Scardac V., Ferrar A.C. and Colemen; *J. Nature Nanotechnology*, Vol. 3, No. 9, p. 538, 2008
100. Hess W. M. and Ban L. L.; *Proc. 6th Int. Congr. on Electron Microscopy* (Kyoto), Vol 1, p 569, 1966
101. Hirata, M., Gotou, T., Horiuchi, S., Fujiwara, M. and Ohba, M. ; *Carbon,* Vol. 42, p. 2929, 2004
102. Hiroshima H. and Komuro M.; *Jpn J Appl Phys*, Vol. 46, p. 6391, 2007.
103. Hiroshima H.; *Jpn J Appl Phys.*, Vol. 47, p. 5151, 2008
104. Hiroshima H., Atobe H., Wang Q. and Youn S-W.; *Jpn J Appl Phys*, Vol. 49, p. 06GL01, 2010.
105. Hsu Pin-Che and Huan-Tsung Chang; *Chem. Commun.*, Vol.48, p. 3984, 2012
106. Hu S.-L., K.-Y. Niu, J. Sun, J. Yang, N.-Q. Zhao and Du X.-W.; *J. Mater. Chem.*, Vol. 19 p. 484, 2009.
107. Hu W.B., Peng C. and Luo W.J.; *ACS Nano.,* Vol. 4, p. 4317, 2010
108. Huang P, Lin J., Wang X, Wang Z., Zhang C, He M., Wang K., Chen F., Li Z., Shen G., Cui D. and Chen X.; *Adv. Mater.,* Vol. 24, p. 5104, 2012
109. Huang L.F., Zhang G. R., Zheng X. H., Gong P. L., Cao T. F. and Zeng Z.; *J. Phys. Condense. Matter,* Vol. 25, No,5, p. 055304, 2013

110. Hummers W. S. and Offeman R. E.; *J. Am. Chem. Soc.*, Vol. 80, No. 6, p. 1339, 1958
111. Ishigami M., Chen J.H., Cullen W.G., Fuhrer M.S. and Williams E.D.; *Nano Lett.*, Vol. 7, p. 1643, 2007
112. Jain N., Bansal T, Durcan C.A., Yang X. and Yu B.; *Carbon*, p. 396, 2013
113. Jain N, Durcan C.A., Jacobs-Gedrim R, Xu Y. and Yu B., *Nanotechnology*, Vol. 24, 355202 2013, doi:10.1088/0957-4484/24/35/355202
114. Jang H.J., Kim Y.K., Kwon H.M., Yeo W.S., Kim D.E. and Min D.H.; *Angew. Chem. Int. Ed.*, Vol. 49, p. 5703, 2010
115. Janowska I., Ovidiu E., Timo J., P., Bégin D., Marc-Jacques Ledoux and Cuong Pham-Huu; *Applied Catalysis A: General;* Vol. 371, No. 1–2, p. 22, 2009
116. Jiang De-en, Mao-Hua Du and Dai S.; *The J. Chem. Phys.*; Vol. 130, No.7, p. 074705, 2009
117. Jiang W., Liu H., Ding Y., Shi Y., Yin L. and Lu B.; *Microelectron Eng.*, Vol. 86, p. 2412, 2009
118. Jiang Z., Pei B., Manthiram A.; *J. Mater. Chem. A*, Vol. 1, p. 7775, 2013
119. Jiang Zhong-jie, Jiang Zhongqing, Chen W.; *Journal of Power Sources*, Vol. 251, p. 55e65, 2014
120. Jiao L., Zhang L., Wang X., Diankov G. and Dai H.; *Nature,* Vol. 458, No.16, p.878, 2009
121. Jingzhi Shang, Lin Ma, Jiewei Li, Wei Ai, Ting Yu and Gurzadyan G.G.; *Sci. Rep.,* Vol. 2, p. 792, 2012
122. Jo G., Chae G.S., Hwang Y.S., Kwon O.N., Lee K.M., Baek K.J. and Rhee T.H; US Patent, 6,881,679 (2005).
123. Joaquim C.G., da Silva E. and Gonçalves H. M.R.; *Trends in Analytical Chem.* Vol. 30, No. 8, p.1327, 2011.
124. Joel Leffler, *Towards graphene based transparent conductive coating*; Masters Thesis, Luleå University of Technology, 2013
125. Juang Z.Y, Wu C.U., Lu A.Y., Su C.Y., Leou K.C., Chen F.R. and Tsai C.H.; *Carbon,* Vol. 48, p. 1088, 2010
126. Juan P., Gao W., Zheng, Liu B.K., Romero-Aburto R., Liehui G., Li Song, Alemany L. B., Zhan X., Gao G., Sajna Antony Vithayathil, Abraham Kaipparettu B., Marti A. A., Hayashi T., Jun-Jie Zhu and Ajayan P. M.; *Nano Lett.* Vol. 12, p.844, 2012
127. Jung J., Pereg-Barnea T and MacDonald A. H.; *Physical Rev. Lett.*; Vol. 102, No.22, p. 7205, 2009
128. Kalita G., Wakita K. and Umeno M.; *Physica E,* Vol. 43, p. 1490, 2011
129. Kalita G., Masahiro M., Uchida H., Wakita K. and Umeno M.; *Materials Lett.*, Vol. 64, p. 2180, 2010
130. Kamat P.V. ; *Chem Rev.*, Vol. 93, N0.1, p.267, 1993
131. Kanad Maharshi; *Vaisesika Sutra*—6th Century BC [Full Text at archive.org of "The Vaisesika sutras of Kanada. Translated by Nandalal Sinha"]
132. Katsnelson M. I.; *Mater. Today*, Vol. 10, p. 20, 2006

133. Kedzierski J., Hsu P.-L., Reina A., Kong J., Healey P., Wyatt P. and Keast C.; *Electron Device Letters, IEEE*, Vol. 30, No.7, p, 745, 2009
134. Khaliq A,R, Raed K., Hamzah M. S. and Waleed F.F.; *Nanotechnology*, Vol. 23, No. 45, p. 455106, 2012
135. Khare R, Mielke S.L., Paci J.T., Zhang S., Ballarini R., Schatz G.C. and Belytschko T; *Phys. Rev. B*, Vol. 75, p. 075412, 2007
136. Kim P., Shi L., Majumdar A. and McEuen P.L.; *Phys Rev. Lett* Vol.87, p 215502, 2001
137. Kim J-G., Sim Y., Cho Y., Seo J-W., Kwon S., Park J-W., Choi H-G., Kim H. and Lee S.; *Microelectron Eng.*, Vol. 86, p. 2427, 2009
138. Kim K. S., Zhao Y., Jang H., Lee S. Y., Kim J. M., Kim K. S. Ahn J. H. Kim P., Choi J. Y. and Hong B. H.; *Nature*, Vol. 457, p. 706, 2009
139. Kobayashi T, Bando M., Kimura N., Shimizu K., Kadono K, Umezu N.,Miyahara K., Hayazaki S., Nagai S., Mizuguchi Y., Murakami Y. and Hobara D.; *Appld. Phys Lett.*, Vol. 102, p. 023112, 2013
140. Kohlschütter V. and Haenni P.; Zeitschrift für anorganische und allgemeine Chemie (in German), Vol. 105, No. 1, p. 144, 1919, doi:10.1002/zaac.19191050109
141. Konwer S., Gogoi J.P., Kalita A. and Dolui S.K.; *J. Materials Sc. Materials in Electronics*, Vol. 22, No. 8, p. 1154, 2011
142. Kosynkin D.V., Higginbotham A.L., Sinitskii A., Lomeda J.R., Dimiev A., Price B., Tour K. and James M.; *Nature*, Vol. 458, No.7240, p. 872, 2009
143. Kosynkin D.V., Lu W, Sinitskii A., Pera G., Sun Z. and Tour J.M.; *ACS Nano*, Vol. 5, p. 968, 2011
144. Krishnan A., Dujardin E., Treacy M.M.J., Hugdahl J., Lynum S. and Ebbesen T.W.; *Nature*, Vol. 388, p. 451, 1997
145. Kroto H.W., Heath J. R., O'Brien S. C, Curl R. F. and Smalley R. E.; *Nature*, Vol. 318, p. 162, 1985.
146. Kwon W. and Rhee S. W.; *Chem Commun. (Camb)* Vol. 48, p. 5256, 2012
147. Lalwani G, Henslee F.B, Liangjun L., Kurtis Kasper F., Yi-Xian Qin, Mikos, A.G. and Sitharaman B.; *Biomacromolecules.* (February/13/2013). doi:10.1021/bm301995s.
148. Lan S., Lee H., Ni J. and Lee M.; *International Conference on Smart Manufacturing Application, 2008* (ICSMA 2008) April 9–11 2008; Gyeonggi-do. Korea: IEEE; p. 371, 2008
149. Land T. A., Michely T., Behm R.J., Hemminger J.C. and Consa G.; *Surf. Sci.*, Vol. 264, p. 261, 1992
150. Landau L.; Z. *Phys.*, Vol. 64, p. 629, 1930
151. Landon O., Westover A., Mares J.W., Chatterjee S., Erwin W.R. Bardhan R., Weiss S.M. and Pint C.L.; *Nature Scientific Reports*, Vol. 3, Article number: 3020; doi:10.1038/srep03020
152. Lee C., Wei X., Kysar J.W. and Hone J.; *Science*, Vol. 321, No. 5887, p. 385, 2008

153. Lee Jae-Ung, Duhee Yoon and Hyeonsik Cheong; *Estimation of Young's Modulus of Graphene by Raman Spectroscopy,* Department of Physics, Sogang University, Seoul 121–742, Korea
154. Lee Mi-Sun, Lee K., So-Yun Kim, Lee H., Park J., Kwang-Hyuk Choi, Han-Ki Kim, Dae-Gon Kim, Dae-Young Lee, SungWoo Nam and Jang-Ung Park; *Nano Lett.*, Vol. 13, p. 2814, 2013
155. Lee S.W., Mattevi C., Chhowalla M. and Sankaran R.M.; *The Journal of Phys. Chem, Lett.*, Vol. 3, p. 772, 2012,
156. Lehtinen P. O., Foster A. S., Ma Y., Krasheninnikov A. V. and Nieminen R. M.; *Phys. Rev. Lett.*, Vol.93, p. 187202, 2004
157. Lerf A., He H., Riedl T, Forster M and Klinowski J.; *Solid State Ionics*, Vol. 857, p.101,1997
158. Lerf A., He H., Forster M and Klinowski J. ; *J. Phys. Chem. B,* Vol. 102, p. 4477, 1998
159. Levallois Julien, Andrew L. Walter, Markus Ostler, Aaron Bostwick, Eli Rotenberg, Thomas Seyller, Dirk van der Marel and Alexey B. Kuzmenko; *Nature Physics,* Vol. 7, p. 48, 2011
160. Li D., Mu¨ller M.B., Gilje S., Kaner R.B. and Wallace G.G.; *Nature Nanotechnol,* Vol. 3, No. 2, p. 101, 2008
161. Li H.T, He X. D, Kang Z. H., Huang H, Liu Y., Liu J.L., Lian S.Y., Tsang C. H. A., Yang X. B. and Lee S. T.; *Angew Chem lnt Ed*, Vol. 49, No.26, p.4430, 2010
162. Li H.T., Kang Z.H., Liu Y., Lee S-T, *J Mater Chem,* Vol. 22, p. 24230, 2012
163. Li N., Wang Z., Zhao K., Shi Z., Gu Z., Xu S., *Carbon*, Vol. 48, p. 255–259, 2010
164. Li X., Zhang G., Bai X., Sun X., Wang X., Wang E.and H. Dai, *Nat. Nanotechnol.,* Vol. 3, p. 538, 2008
165. Li X., Wang X., Zhang L., Lee S. and Dai H.; *Science,* Vol. 319, p. 1229, 2008
166. Li X., Zhu Y., Cai W., Borysiak M., Han B. and Chen D; *Nano Lett.,* Vol. 9, N0. 12, p. 4359, 2009,
167. Li X., Cai W., An J., Kim S., Nah J., Yang D., Piner R., Velamakanni A., Jung I., Tutuc E. *et al.* ; *Science,* Vol. 324, p. 1312, 2009
168. Li Xia and Sun Xueliang, Frontiers in Energy Research, Volume 2, Article 49, 2014, doi: 10.3389/fenrg.2014.00049
169. Li X., Zhu H., Wang K., Cao A, Wei J., Li C., Jia Y., Li Z., Li X. and Wu D.; *Adv. Mater.*, Vol. 22, p, 2743–2748, 2010
170. Liang X., Sperling B.A., Calizo I., Cheng G., Hacker C.A., Zhang Q., Obeng Y., Yan K., Peng H., Li Q., Zhu X., Yuan H., Walker A.R.H., Liu, Z. Peng L.M. and Richter C.A.; *ACS Nano*, Vol.5, No. 11, p. 9144, 2011
171. Liao L., Bai J.W., Lin Y.C., Qu Y.Q., Huang Y., Duan X.F.; *Adv. Materials*, Vol. 22, No. 17, p. 1941, 2010
172. Liao L., Bai J., Qu Y., Lin Y.C., Li Y., Huang Y., Duan X.; *PNAS*, Vol. 107, No. 15, p. 6711, 2010

173. Liao L., Yung-Chen Lin, Mingqiang Bao, Rui Cheng, Jingwei Bai, Yuan Liu, Yongquan Qu, Kang L. Wang, Yu Huang and Xiangfeng Duan; *Nature*, Vol. 467, p. 305, 2010
174. Libin Tang, Rongbin Ji, Xiangke Cao, Jingyu Lin, Hongxing Jiang, Xueming Li,Kar Seng Teng, Chi Man Luk, Songjun Zeng, Jianhua Hao and Shu Ping Lau; *ACS Nano,* Vol. 6, No. 6, p. 5102, 2012
175. Lim H., Choi K-B., Kim G., Park S., Ryu J. and Lee J,; *Microelectron Eng.*, Vol. 88, p. 2017, 2011
176. Lin Y. M., Dimitrakopoulos C., Jenkins K.A., Farmer D.B., Chiu H.-Y., Grill A. and Avouris Ph.; *Science*, Vol. 327, p. 662, 2010
177. Lin Y., Li X., Xie D., Feng T., Chen Y., Song R., Tian H., Ren T., Zhong M., Wan K. and Hongwei Zhu; *Energy Environ. Sci.*, Vol. 6, p. 108, 2013
178. Liu H., Ye T., Mao C.; *Angew Chem. Int. Ed*, Vol. 46 p. 6473, 2007
179. Liu M.Z., Robinson J.T., Sun X.M. and Dai H.J.; *J. Am. Chem. Soc.*, Vol.130, p. 10876, 2008
180. Liu Y., Dingshan Yu, Chao Zeng, Zongcheng Miao and Liming Dai; *Langmuir,* Vol. 26, No. 9, p. 6158, 2010
181. Liu R., Wu D., Feng X. and Muellen K.; *J Am Chem Soc.*, Vol. 133, p. 15221, 2011
182. Liu S., Tian J.Q., Wang L., Zhang Y.W., Qin X.Y., Luo Y.L., *et al.*; *Adv Mater;* Vol.24, No.15, p.2037, 2012
183. Liu L., Park J., Siegel D.A., McCarty K.F. K.W., W., Basile L., Idrobo C.J., Li A-P and Gu G.; *Science*, Vol. 343. No. 6167, p. 163, 2014
184. Liying J., Zhang L., Wang X., Diankov G. and Dai H.; *Nature Letters.*, Vol. 458, 2009
185. Loh K. P., Bao Q., Eda G. and Chhowalla M.; *Nature Chem.* Vol. 2, p. 1015, 2010
186. Longchamp J-N., Latychevskaia T., Escher C. and Hans-Werner Fink; *Physical Rev Letters*, Vol.110, No. 25, p. 255501, 2013
187. Lotya M., Hernandez Y., King P.J., Smith R.J., Nicolosi V., Karlsson L.S., Blighe F.M., De S., Wang Z.M., McGovern I.T., *et al.*; *J.Am. Chem. Soc.*; Vol. 131, p. 3611, 2009
188. Lu C.H., Yang H.H., Zhu C.L., Chen X. and Chen G.N.; *Angew. Chem. Int. Ed.*, Vol. 48, p. 4785, 2009
189. Lui C.H., Liu L., Mak K.F., Flynn G.W. and Heinz T.F.; *Nature*, Vol. 462, p. 339, 2009
190. Luican A, Li G. and Andrei E.Y., arXiv: 0803.4016. Department of Physics and Astronomy, Rutgers University, New Jersey, USA, 2008
191. Ma L., Wang J. and Feng D.; *ChemPhysChem*, Vol.14, No 1, p. 47, 2013
192. Magaud L., Hiebel F., Varchon F., Mallet P. and. Veuillen J.-Y; *Rapid Research Letters,* Vol. 3, p. 172, 2009
193. Mäkelä T., Haatainen T., Majander P. and Ahopelto J.; *Microelectron. Eng,* Vol. 84, p. 877, 2007

194. Mäkelä T., Haatainen T. and Ahopelto J.; *Microelectron. Eng.*, Vol. 88, p. 2045, 2011
195. Malard L.M., Pimenta M.A., Dresselhaus G., Dresselhaus M.S.; *Physics Reports*, Vol. 473, p. 51, 2009
196. Mao X.J., Zheng H.Z., Long Y.J., Du J., Hao J.Y., Wang L.L. and Zhou D.B.; *Spectrochim. Acta Part A*, Vol. 75, p. 553, 2010
197. Marchini S., Günther S. and Wintterlin J.; *Phys. Rev. B*, Vol.76, p. 075429, 2007
198. Marconi U. M. B., Puglisi A., Rondoni L. and Vulpiani A.; *Phys. Rep.* Vol. 461, p. 111, 2008
199. Markovic Z.M., Harhaji-Trajkovic L.M., Todorovic-Markovic B.M. *et al.*; *Biomaterials*, Vol. 32, p. 1121, 2010
200. Martin Luiz G. P, Yi Song, Tingying Zengb, Mildred S., Dresselhaus M.S., Kong J. and Araujo P.T.; *Proc Natl Acad Sci U S A.*, Vol. 110, No. 44, p. 17762, 2013
201. McAllister M.J., Li J-L., Adamson D.H., Schniepp H.C., Abdala A.A., Liu J., *et al*; *Chem Mater*; Vol. 19, No. 18, p. 4396, 2007
202. Merino S., Retolaza A., Juarros A. and Landis S.; *Microelectron. Eng.*, Vol. 84, p. 848, 2007
203. Mewada A., Pandey S., Thakur M., Jadhav D. and Madhuri Sharon; *J. Mater. Chem. B*, Vol. 2, p. 698, 2014, DOI: 10.1039/C3TB21436
204. Mewada A, Pandey S. Shinde S., Mishra N, Oza G, Thakur M., M. Sharon and Madhuri Sharon; *Mater. Sci. Eng. C*, Vol. 33, No. 7, p.3716, 2013
205. Meyer J.C., Geim A.K., Katsnelon M.I., Novoselov K.S., Booth T.J. and Roth S.; *Nature*, Vol. 446, No. 7131, p. 60, 2007
206. Meyer Jannik C, *Nano Lett.*, Vol. 8, No. 11, p. 3582, 2008
207. Miao C., Zheng C., Liang O. and Xie Y-H.;. Chemical Vapor Deposition of Graphene, INTECH Open Access Publisher, 2011.
208. Shin M.K., Lee B., Kim S.H., Lee J.H., Spinks G.M., Gambhir S., Wallace G.S., Kozlov M.E., Baughman R.H. and Kim S.J.; *Nat Commun*; Vol. 3, p. 650, 2012
209. Ming H., Ma Z., Liu Y., Pan K.M., Yu H., Wang F. and Kang Z.H.; *Dalton Trans.* Vol. 41, p. 9526, 2012.
210. Mkhoyan K. A., Contryman A. W., Silcox J., Stewart D. A., Eda G., Mattevi C., Miller S. and Chhowalla M.; *Nano Lett.*, Vol. 9, p. 1058, 2009
211. Mohamed K.; *PhD Thesis*, University of Canterbury, Electrical and Computer Engineering, 2009
212. Mohanty N. and Berry V.; *Nano Letters*, Vol. 8, No. 12, p. 4469, 2008
213. Mohanty N., Moore D., Zhiping Xu, Sreeprasad T.S., Ashvin Nagaraja, Rodriguez A. A. and Berry V; *Nature Communications*, Vol. 3, p. 844, 2012
214. Morozov S. V., K. S., M. I., F., Elias D.C., J.A., and Geim A.K.; *Physical Review Letters*, Vol. 100, p. 016602, 2008
215. Morozov S. V., Novoselov K. S, Schedin F., Jiang D., Firsov A. A. and Geim A. K.; *Phys. Rev. B*, Vol. 72, p. 201401, 2005

216. Mouras S., Hamm A., Djurado D. and Cousseins J. C.; *Revue de Chimie Minerale,* Vol. 24, p. 572, 1987
217. Mukhopadhyay P. (2013). *Graphite, Graphene and their Polymer Nanocomposites*. Boca Raton, Florida: Taylor and Francis Group. p. 202–213. ISBN 978-1-4398-2779-6
218. Mukhopadhyay P. and Gupta R. edited, Epoxy Nanocomposites—Graphene promising filler in *Graphene Based Polymer Nanocomposites,* CRC Press, 2013
219. Murata Y., Petrova V., Kappes B.B., Ebnonnasir A., Petrov I., Xie Y-H., Ciobanu C.V. and Kodambaka S.; *ACS Nano,* Vol. 4, No. 11, p. 6509, 2010
220. Mustafa L., King P.J., Khan U., De S. and Coleman J.N.; *ACS Nano,* Vol. 4, No. 6, p. 3155, 2010
221. Naebe M., Wang J., Abbas A., Khayyam H., Hameed N., Li L.H., Chen Y. and Fox B.; *Scientific Reports,* Vol. 4, Article No. 4375 doi:10.1038/srep04375
222. Nagashima A., Nuka, K., Satoh, K.; Itoh H. and Ichinokawa, *Surf. Sci.,* Vol. 291, p. 93, 1993
223. Nagato K., Sugimoto S., Hamaguchi T. and Nakao M.; *Microelectron. Eng.,* Vol. 87, p. 1543, 2010
224. Nair R.R., Tsai I.-L., Sepioni M., Lehtinen O., Keinonen J., Krasheninnikov A.V., Castro Neto A.H., Katsnelson M.I., Geim A.K. A.K. and Grigorieva I.V., *Nature Communications.* 2013 June; 4: eScholarID:205884 | DOI:10.1038/ncomms3010
225. Nair R. R., Wu H. A., Jayaram P. N., Grigorieva I. V. and Geim A. K.; *Science,* Vol. 335, No. 6067, p. 442, 2012
226. Nakada K., Fujita M., Dresselhaus G. and Dresselhaus M.S.; *Physical Review B,* Vol. 54, No. 24, 1996
227. Natalya S. A., Stroyuk O.L., Yanchuk I.B. and Yefanov A.V.; *Colloid. Polym. Sci.,* Vol. 292,p. 539, 2014
228. Nomura K. and A.H.; *Phys. Rev. Lett.,* Vol. 96, Article No. 256602
229. Nawaz F., Wang L., Zhu Long-Feng, Meng Xiang-ju, Xiao Feng-Shou.; *Chemical Research in Chinese Universities,* Vol. 29, No.3, p. 401, 2013
230. Nelson T., Zhang B. and Prezhdo O.V.; *Nano Lett.,* Vol. 10, p. 3237, 2010
231. Neto A.H.C., Guinea F., Peres N.M.R., Novoselov K.S. and Gein A.K.; *Rev. Mod. Phys.; Vol.* 81, p. 109, 2009
232. Ng Y.H., Iwase A., Kudo A. and Amal R.; *J Phys Chem Lett.,* Vol. 1, No. 17, p. 2607, 2010
233. Ni Guang-Xin, Yi Zheng, Sukang Bae, Chin Yaw Tan, Orhan Kahya, Jing Wu, Byung Hee, Hong, Kui Yao and Barbaros Özyilmaz; *ACS Nano,* Vol. 6, No. 5, p. 3935, 2012
234. Nika D.L., Pokatilov E.P., Askerov A.S., Balandin A.A.; *Phys. Rev.,* Vol.79, No. 155413, 2009
235. Novoselov K. S., Geim A. K., Morozov S. V., Jiang D., Zhang Y., Dubonos S. V., Grigorieva I. V. and Firsov A. A. *Science,* Vol. 306, No. 666, 2004

236. Novoselov K. S., Geim A. K., Morozov S. V., Jiang D., Katsnelson M.I, Grigorieva I.V., Dubonos S. V. and Firsov A. A.; *Nature,* Vol. 438, p. 197,2005
237. Novoselov K. S., Jiang D., Schedin F., Booth J., Khotkevich V.V., Morozov S.V. and Geim A.K.; *Proceedings of the National Academy of Sciences of the United States of America,* Vol.102, No. 30, p. 10451, 2005
238. Novoselov K.S., McCann E.,. Morozov S. V, Fal'ko V. I., Katsnelson M. I.,. Zeitler U,. Jiang D, Schedin F. and Geim A. K.; *Nature Physics*, Vol. 2, p. 177, 2006
239. Ohta T., Gabaly E.I., Bostwick F., McChesney A., Emtsev J.L., Schmid K.V., Andreas K., Seyler T., Karsten H and Rotenberg E. ; *New Jour. Phys.*, Vol. 10, p. 023034, 2008
240. Ohta T., Bostwick A., Seyller T., Horn K. and Rotenberg E.; *Science*, Vol. 313, p. 951, 2006
241. Omeltchenko A., Yu J., Kalia R.K. and Vashishta P.; *Phys. Rev. Lett.*, Vol. 78, p. 2148, 1997
242. Oshima, C.; Nagashima, A.; *J. Phys.: Condens. Matter*, Vol. 9, p. 1, 1997
243. Pan G., Bing Li, Heath M., HorsellD,. Wears M.L., Laith Al Taan and Shakil A,; *Carbon*, Vol.65, p. 349-358, 2013
244. Pan Y., Shi D. X. and Gao H. J.; *Chin. Phys.*, Vol. 16, p. 3151, 2007
245. Pan Y., Jiang N., Sun J. T., Shi D. X., Du S. X., Liu F. and Gao H.-J.; *Adv. Mater.,* Doi:1002/adma.200800761; arXiv:0709.2858v1
246. Pandey S., Thakur M., Ashmi M., D., Mishra N. and Madhuri Sharon; *RSC Advances*, 2013, DOI:10.1039/C3RA42139B
247. Pandey S., Mewada A, Thakur M., Pillai S., Dharmatti R., Phadke C. and Madhuri Sharon; *RSC Advances* 10/2013; DOI:10.1039/C3RA42139B]
248. Park, S., Lee, K.-S., Bozoklu, G., Cai, W., Nguyen, S.T. and Ruoff, R.S.; *ACS Nano,* Vol. 2, p. 572, 2008
249. Park S., Choi K., Kim G. and Lee J.; *Microelectron Eng.*, Vol. 86. p. 604, 2009
250. Park H. and Ruoff R. S.; *Nat. Nanotechnol.*, Vol. 4, p. 217, 2009
251. Park H. J., Meyer J., Roth S. and Skákalová V.; *Carbon,* Vol. 48, No. 4, p. 1088, 2010
252. Park H., Chang S., Smith M., Gradečak S., Kong J.; *Sci Rep.*, Vol. 3, p. 1581, 2013
253. Park H., Chang S., Jean J., Cheng J.J., Araujo P.T., Wang M., Bawendi M.G., Dresselhaus M.S., Bulović V., Kong J. and Gradečak S.; *Nano Lett.*, Vol. 13, No. 1, p. 233, 2013
254. Partoens B. and Peeters F.M.; *Phys. Rev. B*, Vol. 74, p. 075404, 2006
255. Parvez K., Yang S., Hernandez Y., Winter A., Turchanin A, Feng X. and Mullen; *AC Nano,* Vol. 6, No. 11, p. 9541, 2012
256. Patel P. ; *MIT Technology Review,* 2009
257. Pei S., Zhao J., Du J, Ren W. and Cheng H-M.; *Carbon,* Vol. 48, p. 4466, 2010
258. Peigney A., Laurent C., Flahaut E., Bacsa R. R. and Rousset A.; *Carbon*, Vol. 39, p.507,2001

259. Peng, H. and Travas-Sejdic, J.; *Chemistry of Materials*. Vol. 21, No. 23, p. 5563, 2009.
260. Peng, J., Gao W., Gupta B.K., Liu Z., Rebeca Romero-Aburto, Liehui Ge, Song L., Alemany L.B., Zhan X., Gao G., Vithayathil S.A., Kaipparettu B.A., Marti A.A., Hayashi T., Zhu J.J. and Ajayan P.M.; *Nano Lett*. Vol. 12, p. 844, 2012
261. Pop E., Mann D., Wang Q., Goodson K. and Dai H.; *Nano Lett*, Vol. 6, p. 96, 2005
262. Prakash G., Capano M.A., Bolen M.L., Dmitry Z. and Reifenberger R.G.; *Carbon*, Vol. 48, No. 9, p. 2383, 2010
263. Putz K.W., Compton O.C., Palmeri M.J., Nguyen S.T.and Brinson L.C.; *Adv. Funct. Mater.*, Vol. *20*, p. 3322, 2010
264. Puvvada N., Prashanth Kumar B.N., Konar S., Kalita H., Mandal M. and Patha A.; *Sci. Technol. Adv. Mater.* Vol. 13, No. 4, p. 045008.
265. Qi L., Zhu S., Luo X., Zou M. and Huang S.; *AIP Advances*, Vol.2, p.032146, 2012.
266. Qi X., Tan C., Wei J. and Zhang H. ; *Nanoscale*, Vol. 5, p. 1440, 2013
267. Qinlong Wang, Xiaoxiao H., Yijuan L., Wang X., Zhang, H., Zhu R., Liping L., Ping T. and Zheng H.; *Carbon*, Vol. 59, p.192 2013
268. Qu K. G., J. S. Wang, Ren J. S. and. Qu X. G.; *Chem. Eur. J.*, Vol. 19, No.22, p. 7243, 2013
269. Qu L., Liu Y., Baek J.-B. and Dai L.; *ACS Nano*, Vol. 4, p. 1321, 2010
270. Qu L., Zhang Z., Zhang J. and Chen N.*; Energy Environ. Sci.* Vol. 5, p. 8869, 2012
271. Que Yande Wende Xiao, Xiangmin Fei, Hui Chen, Li Huang, S. X. Du and H.-J. Gao; *Adv. Mater.*, Vol. 21, p. 2777, 2009
272. Qiong Wu,; William Regan,; Hsin-Zon Tsai, Alex Zettl and; Michael F. Crommiel; *Nature Nanotechnology*, Vol. 7, No. 7, published online 20 July 2008
273. Raifee M., Wei L., Thomas A.V., Zandiatashbar A., Rafiee J. and Tour J.M.; *ACS Nano*, Vol. 4, No.12, p. 7415, 2010
274. Ramanathan T., Abdala A. A., Stankovich S., Dikin D.A., Herrera-Alonso M., Piner R.D., Adamson D.H., Schniepp H.C., Chen X., Ruoff R.S., Nguyen S.T., Aksay I.A., Prud'Homme R.K. and Brinson L.C.; *Nat. Nanotechnol.*, Vol. 3, p. 327, 2008
275. Ramesha G.K. and Sampath S.; *J Phys Chem C*, Vol. 113, No.9, p. 7985, 2009
276. Rasheed A. K., Raed M. K., Waleed F.F. and Hamzah M.S.; *Proceedings of the 2nd International Conference on Biotechnology Engineering ICBioE'11*, May 17-19, 2011, Kuala Lumpur, Malaysia
277. Ray S.C., Saha A., Jana N.R. and Sarkar R., *J Phys Chem C*, Vol. 113, No. 43, p. 18546, 2009
278. Reina A., Jia X., Ho J., Nezich D., Son H., Bulovic V., Dresselhaus M.S. and Kong J.; *Nano Lett.*, Vol. 9, No.1, p. 30, 2009
279. Ruammaitree A., Nakahara H., Akimoto K., Soda K. and Saito Y., *Science*, Vol. 282 p. 297, 2013

280. Ruess G. and Vogt F.; *Monatshefte für Chemie*; Vol. 78, No. 3–4, p. 222, 1948
281. Ruffieux P., Gröning O., Schwaller P., Schlapbach L. and Gröning P.; *Phys. Rev. Lett.*, Vol. 84, p. 4910, 2000
282. Salehi-Khojin A., Estrada D., Lin K.Y., Bae M-H., Xiong F, Pop E. and Masel R.I.; *Advanced Materials*, Vol. 24, No.1, p. 53, 2012
283. Salinas-Castillo A., Ariza-Avidad M., Pritz C, Camprubí-Robles M, Fernández B., Ruedas-Rama M.J., Fernández A.M., Lapresta Fernández A., Santoyo-Gonzalez F., Schrott-Fischer A. and Capitan-Vallvey L. F.; *Chem. Commun.*, Vol 49, p.1103, 2013
284. Sampath S.K.; *J Phys Chem C*, Vol.13, No. 19, p. 7985, 2009.
285. Satya Kishore M., Srimathi P., Kumar S., Swarnagowri A., Srinivasan S., Tilak V and Colborn R.; *Science*, Vol. 321, p. 385, 2008;
286. Schedin, F.; Geim, A. K., Morozov, S. V., Hill, E. W., Blake, P., Katsnelson, M. I. and Novoselov, K. S.; *Nature Materials,* Vol. 6, No. 9, p. 652, 2007
287. Schniepp H.C., Li J-L., McAllister M.J., Sai H, Herrera-Alonso M., Adamson D.H., Prud'homme R.K., Car R., Saville D.A, and Aksay I.A. ; *J Phys Chem B*, Vol.110, No. 17, p.8535, 2006
288. Scott J. B., Scott S. V., Alden J.S., van der Zande A.M., Parpia J.M., Craighead H.G. and McEuen P.L.; *NANO Lett.,* Vol.8, No... 8, p. 2458, 2008
289. Secor E.B., Prabhumirashi P.L., Puntambekar K., Geier M.L. and Hersam M.C.; *J. Phys. Chem. Lett.*, Vol. 4, No. 8, p 1347, 2013
290. Segal M. ; *Nature Nanotechnol.*, Vol.4, p. 612, 2009
291. Semenoff. Gordon W.; *Phys. Rev. Lett.*, Vol. 53, p. 2449, 1984
292. Shan C.S., Yang H.F., Song J.F., Han D.X., Ivaska A., Niu L.; *Anal. Chem.*, Vol. 81, p. 2378, 2009
293. Shang X., Zhang M., Wang X. and Yang Y.; *Jour. of Exptl. Nanoscience*; Vol. 9, No.7, p. 749, 2014
294. Shang J., Lin Ma, Jiewei Li, Wei Ai, Ting Yu and Gagik G. Gurzadyan; *Scientific Reports,* Vol. 2,Article No.792 doi:10.1038/srep00792
295. Shao Y.Y., Wang J., Wu H., Liu J., Aksay I.A. and Lin Y.H.; *Electroanalysis,* Vol. 22, p. 1027, 2010
296. Sharon Maheshwar, Jain S., Kichambre P.D., Kumar M., *Materials Chemistry, and Physics*, Vol. 8, p. 284, 1998
297. Shi Y., Kim K.K., Reina A., Hofmann M., Li L.J. and Kong J.; *ACS Nano*, Vol. 4, No. 5, p. 2689, 2010
298. Shin H-J, Kim KK, Benayad A, Yoon S-M, Park HK, Jung I-S; *Adv. Funct Mater*, Vol. 19, No. 12, p.1987, 2009
299. Shinde D.B. and Pillai V.K.; *Chemistry—A European Journal,* Vol. 18, No.39, p. 12522, 2012
300. Shinde D.B., Majumder M. and Pillai V.K.; *Scientific Reports*, Vol.4, p. 4363, 2014
301. Shinohara H., Fukuhara M., Hirasawa T., Mizuno J. and Shoji S.; *J Photopolymer Sci. Technol.,* Vol. 21, p. 591, 2008,

302. Singh R.S., Venkatram N., Wei C., Wee A., Thye S. and Wei J.; *ACS Nano*, Vol.5, No. 7, p.5969, 2011
303. Singh R.K., Raman P., Banerjee C., Derrek E. Lobo, Gullapalli H., Sumandasa M., Kumar A., Choudhary L., Tkacz R., Ajayan P.M. and Majumder M.; *Carbon*, Vol. 50, Issue 11, p. 4040, 2012
304. Sinitskii A., Fursina A. A., Kosynkin D. V., Higginbotham A. L., Natelson D., and Tour J. M. ; *Appl. Phys. Lett.*, Vol. 95, p. 253108, 2009
305. Sinitskii A., Fursina A. A., Kosynkin D. V., Higginbotham A. L., Natelson D., Pei S., Zhao J., Du J., Ren W. and Hui-Ming Cheng; *Carbon*, Vol. 48, p. 4466, 2010
306. Shioyama M., Kawanishi T., Yokoyama S., Nuno M. and Yamamoto T.; *Water Supply*, Vol. 1, No. 5, p. 91, 2001
307. Sohn K-J., Park J.H., Lee D-E., Jang H-I. and Lee W.I.; *J Micromech Microeng*, Vol. 23, p. 035024, 2013
308. Son Y.-W., Cohen M. L. and Louie S. G. ; *Phys. Rev. Lett.*, Vol. 97, No.21, 2006
309. Song Y.J., Qu K.G., Zhao C., Ren J.S., Qu X.G., *Adv. Mater.*, Vol. 22, p. 2206, 2010
310. Song S. M., Park J.K., Sul O.J. and Cho B.J., *Nano Lett.*, Vol.12, p. 3887, 2012
311. Srivastava S., Awasthi R., Tripathi D., Rai M. K., Agarwal V., Gajbhiye N.S. and Gupta R. K., *Small*, Vol. 8, p. 1099, 2012
312. Stankovich S. D., Dommett G.H.B., Kohlhaas K.M., Zimney E. J., Stach E.A., Piner R.D., Nguyen S.T and Ruoff R.S.; *Nature*, Vol. 442, p. 282, 2006
313. Stankovich S., Piner R.D., Chen X., Wu N., Nguyen S.T. and Ruoff R.S.; *J Mater Chem.*, Vol. 16, No. 2, p. 155, 2006
314. Stöckle R. M., Suh Y. D., Deckert V. and Zenobi R.; *Chem. Phys. Lett.*, Vol. 318, p. 131, 2000
315. Stoller M.D., Park S., Zhu Y., An J. and Ruoff R.S.; *Nano Letters*, Vol. 8, p. 3498, 2008
316. Subrahmanyam K. S., Panchakarla L. S., Govindaraj A., Rao C. N. R., *J. Phys. Chem. C*, V0l. 113, p. 4257–4259, 2009
317. Sugunan A., Zhao Y, Somak M., Lin D.; Li S., Popov S. and Marcinkevicius S.; *Nanotechnology*, Vol. 22, No. 42, p. 425202, 2011
318. Sun Y. P., Zhou B., Lin Y., Wang W., Fernando K.A.S., Pathak P., Meziani M.J., Harruff K. A., Wang X., Wang H., Luo P.G., Yang H, Kose M.E., Chen B., Veca L.M. and Su-Yuan Xie; *J. Am. Chem. Soc.*, Vol. 128, p. 7756, 2006
319. Sun X., Liu Z., Welsher K., Robinson J.T.,Goodwin A., Zaric S. and Dai H.; *Nano Res.*, Vol. 1, p. 203, 2008
320. Sutter P. W., Flege J. I. and Sutter E. A.; *Nat. Mater.*, Vol. 7, p, 406, 2008
321. Tan C., Rodríguez-López J., Park J.J., Ritzert N.L., Ralph and Héctor D. Abruña ; *ACS Nano*, Vol. 6, No. 4, p. 3070, 2012
322. Tang L.A...L, Wang J.Z. and Loh K.P.; *J. Am. Chem. Soc.*, Vol. 132, p. 10976, 2010
323. Tang L.B., Ji R.B., Cao X.K., Lin J.Y., Jiang H.X., Li X.M., Teng K.S., Luk C.M., Zeng S.J., Hao J.H. and Lau S.P.; *ACS Nano*, Vol. 6, No. 6, p. 5102, 2012

324. Tapasztó L., Dobrik G., Lambin P. and Biró L. P.; *Nature Nanotechnology*, Vol.3, No.7, p. 397, 2008
325. Terdalkar S.S., Huang S., Yuan H., Rencis J.J., Zhu T. and Zhang S.; *Chem. Phys. Lett.*, Vol. 494, p. 218, 2010
326. Teweldebrhan D. and. Balandin A. A.; *Appld. Phys. Let.*, Vol. 94, p. 013101, 2009
327. Thompson Innovation database https://www.thomsoninnovation.com
328. Tian L., Ghosh D., Chen W., Pradhan S., Chang X. and Chen S.; *Chem Mater*, Vol. 21, p. 2803, 2009,
329. Tian B., Wang C., Zhang S., Feng L. and Liu Z.; *ACS Nano,* Vol. 5, p. 7000, 2011
330. Tongay S., Lemaitre M., Miao X., Gila B., Appleton B. R. and Hebard A. F.;*Physical Review,* Vol. X, No. 2. P. 011002, 2012
331. Tour J. M.; *Appl. Phys. Lett.*, Vol. 95, p. 253108, 2009
332. Tsoukleri G., Parthenios J., K., R., A.C., Geim A.K., K.S. and C.; *Small*, Vol 5, p. 2397, 2009
333. Tung V. C., Allen M. J., Yang Y. and Kaner R. B.; *Nat. Nanotechnol.*, Vol.4, p. 25, 2009
334. Tuinstra F. and Koenig L.; *J. Chem. Phys.* Vol. 53, p. 1126, 1970
335. Tran Viet Cuong, Viet Hung Pham, QuangTrung Tran, Jin Suk Chung, Eun Woo Shin, Jae Seong Kim, Eui Jung Kim; *Materials Lett.,* Vol. 64, p. 765, 2010
336. Tzalenchuk A., Lara-Avila S., Kalaboukhov A., Paolillo S., Syväjärvi M., Yakimova R., Kazakova O., Janssen T. J. B. M., Fal'ko V. and Kubatkin S.; *Nature Nanotechnol.*, Vol. 5, p. 186, 2010
337. Ubbelohde A.R., Lewis F.A.; *Graphite and its Crystal Compounds.* Oxford, Clarendon Press. (1960).
338. van Bommel A. J., Crombeen J. E. and van Tooren A.; *Surf. Sci.*, Vol. 48, p. 463, 1975.
339. Vázquez de Parga A. L., Calleja F., Borca B., Passeggi M. C. G., Hinarejos J. J., Guinea F. and Miranda R.; *Phys. Rev. Lett.*, Vol. 100, p. 056807, 2008
340. Vlassak J.J. and Nix W.D.; *J. Material Res.,* Vol. 7, No. 12, p. 3242, 1992
341. Vlassiouk I., Pasquale F., Meyer H., Lavrik N., Dai S., Datskos P. and Smirnov S.; *Carbon,* Vol. 54, p. 58, 2013
342. Wakabayashi K., Fujita M., Ajiki H. and Sigrist M.; *Physical Review B,* Vol. 59, No.12, p. 8271, 1999
343. Wakeland S., Ricardo M., Grey J.K. and Luhrs C.C.; *Carbon*, Vol. 48, No. 12, p. 3463, 2010,
344. Walker S.; *Science Daily*. 24 July 2010.
345. Wallace P. R.; *Phys. Rev.*, Vol. 71, p. 622, 1947
346. Wang Z. F., Shi Q. W., Li Q., Wang X., Hou J. G., Zheng H., Yao Y. and Chen J.; *Applied Physics Letters,* Vol. 91, No. 5, p. 053109, 2007
347. Wang X., Ouyang Y., Li X., Wang H., Guo, J. and Dai H. *Phys Rev.Lett,* Vol. 100, No. 20, 2008a

348. Wang S., Chia P., Chus L.L., Zhao L.H., Png R.Q., Shivaramakrishnan S. *et al.*; *Adv. Mater.*, Vol. 20, p. 3440, 2008
349. Wang Y. Y., Ni Z. H., Yu, T., Wang H. M., Wu Y. H., Chen W., Wee A. T. S. and Shen, Z. X.; *J. Phys. Chem.C*, Vol. 112, p.10637, 2008
350. Wang D., Choi D., Li J., Yang Z., Zimin Nie, Rong Kou, Hu D., Wang C., Saraf L.V., Zhang J., Aksay I.A. and LiuJ.; *ACS Nano,* Vol. 3, No. 4, p. 907, 2009.
351. Wang G., Wang B., Park J., Yang J., Shen X. and Yao J., *Carbon,* Vol.47, p. 68, 2009
352. Wang G., Shen X., Wang B., Yao J. and Park J.; *Carbon*, Vol.47, p.1359, 2009
353. Wang G., Shen X., Wang B., Yao J. and Park J.; *Carbon,* Vol. 47, p. 1359, 2009
354. Wang Y., Li Z.H., Hu D.H., Lin C.T., Li J.H. and Lin Y.H.; *J. Am. Chem. Soc.* Vol. 132, p. 9274, 2010
355. Wang K., Ruan J., Song H., Zhang J, Wo Y., Guo S. and Cui D.; *Nanoscale Res Lett.*, Vol. 6, p. 8, 2011
356. Wang Q., Xiaoxiao H., Yijuan L., Wang X., Zhang H., Rui Zhu, Liping L., Ping Teng and Zheng H.; *Carbon*, Vol. 59, p. 192, 2013
357. Wang D., Yonggang M., Youhai Yu and Bo Peng.; *Jour. Colloid and Interface Sci,* Vol. 417, p. 270, 2014
358. Wang H., Zheng X., Chen H., Yan K., Zhu Z. and Yang S.; *Chem. Commun.*, Vol. 50, p.7517, 2014
359. Wang J. T. W.; Ball, J. M.; Barea, E. M.; Abate, A.; Alexander-Webber, J. A.; Huang, J.; Saliba, M., Mora-Sero I. N., Bisquert J., Snaith H. J. and Nicolas R. J.; *Nano Letters*: 131216131814005, doi:10.1021/nl403997a
360. Wang Z. *et al.*; *Acta Phys. Sin.* Vol. 60, No. 1, p. 017102, 2011
361. Wassei J.K. and Kaner R.B., *Materials today*, Vol. 13, No. 3, p. 52, 2010
362. Waugh R.; *Daily Mail* (London), 2013
363. Wei D. and Liu Y.; *Adv. Mater.*, Vol. 22, p. 3225, 2010
364. Wei H., Pi K., McCreary K. M., Li Y., Wong J. J. I., Swartz A. G. and. Kawakami R. K.; *Phys. Rev. Lett.*, Vol. 105, p. 167202, 2010
365. Wei L.V., Feng Sun, Dai-Ming Tang, Hai-Tao Fang, Chang Liu, Quan-Hong Yang and Hui-Ming Cheng ; *J. Mater. Chem.*, Vol. 21, p. 9014, 2011
366. Wei D., Xie L., K.K., Hu Z., Tan S., Wei C., C.H, K., Y. and A.T.S.; *Nature Communications, Vol.*4, No. 1374, 2012
367. Williams G., Seger B. and Kamat P.V.; *ACS Nano*, Vol. 2, No.7, p. 1487, 2008
368. Williams G. and Kamat P.V.; *Langmuir*, Vol. 25, No. 24, p. 13869, 2009
369. Wonbong Choi, Indranil Lahiri, Raghunandan Seelaboyina andYong Soo Kang; *Solid State and Materials Sciences,* Vol. 35, No. 1, p. 52, 2010
370. Wood D., Schmucker S.W., Lyons A.S, Pop E. and Lyding J.W.; *Nano Letters*, Vol.11, p. 4547, 2011
371. Wood J.D., Schmucker S.W., Raasch R.T., Doidge G.P., Nienhaus L, Damhorst G.L., Lyons A.S., Gruebele M, Bashir R., Pop E. and Lyding J.W., *12th IEEE International Conference on Nanotechnology (IEEE- NANO)* 20–23
372. Wu J. B., Becerril H. A., Bao Z. N., Liu Z. F, Chen Y. S. and Peumans P.; *Appl. Phys.* Lett., Vol. 92, p. 263302, 2008

373. Wu L., M. Luderer, X. Yang, C. Swain, H. Zhang, K. Nelson, A. J. Stacy, B. Shen, G. M. Lanza and D. Pan, *Theranostics,* Vol. 3, p. 677, 2013
374. Wu Z. S., Ren W. C., Gao L. B., Zhao J. P., Chen Z. P., Liu B. L., Tang D. M., Yu B., Jiang C. B. and Cheng H. M.; *ACS Nano,* Vol. 3, p. 411–417.2009
375. Wu Z.S., Ren W., Wang W., Li F., Liu B., and Cheng H.M.; *ACS Nano,* Vol. 4, No. 10, p. 5835, 2010
376. Wu Z.-S., Ren W., Xu L., Li F., Cheng H.-M, *ACS Nano,* Vol. 5, p. 5463e5471, 2011
377. Zhang X., Wang S., Zhu C., Liu M., Jia Y., Lin F., Lei Tao and Wei Y.; *Journal of Colloid and Interface Science*, Vol. 397, p. 39, 2013
378. Xinming Li, Hongwei Zhu, Wang K., Cao A., Wei J., Li C., Yi Jia, Zhen Li, Xiao Li and Wu D.; *Adv. Mater.* Vol. 22, p. 2743, 2010
379. Xiu C.Y., Gao Q., Li G.D., Wang K.X. and Chen J. S.; *Inorg. Chem.,* Vol.13, p. 5859, 2010
380. Xiu-Yun C.; *International Nano Letters, Vol.* 2013,p. 3, 2013
381. Xu K., Huang J. and Li Y.; *Sensor (Basel),* Vol. 9. No. 7. P. 5534, 2009
382. Xu M., Tabarraei A., Paci J.T., Oswald J. and Belytschko T.*; Int. J. Fract.*, Vol. 173, p. 163, 2013
383. Xu X., Ray R., Gu Y., Ploehn H.J., Gearheart L., Raker K. and Scrivens W.A., *J Am Chem Soc,* Vol.126, p. 12736, 2004,
384. Xu Y.X., Bai H., Lu G.W., Li C. and Shi G.Q.; *J Am Chem Soc,* Vol. 130, p. 5856, 2008
385. Xue K. and Xu Z.; *Applied Physics Letters*, Vol. 96, No. 6, p. 063103, 2010
386. Yamada T., Ishihara, M., Kim J., Hasegawa M. and Iijima S.; *Carbon*. Vol. 50, p, 2615, 2012
387. Yan K., Lei Fu, Hailin Peng, and Zhongfan Liu; *Acc. Chem. Res.*, Volume 46, DOI: 10.1021/ar400057n
388. Yang, S. T., Wang, X., Wang, H., Lu, F., Luo, P. G., Cao, L. and Meziani M. J.; *The Jour. Phys. Chem. C, Nanomaterials and interfaces*, Vol.113, No. 42, p. 18110, 2009, doi:10.1021/jp9085969
389. Yang D., Velamakanni A., Bozoklu G., Park S., Stollera M., Piner R.D, Stankovich S., Jung I., Field D.A., Ventric Jr.d C.A. and Ruoff R.S.; *Carbon,* Vol. 47, p. 145, 2009
390. Yang X.Y., Zhang X.Y., Liu Z.F., Ma Y.F., Huang Y. and Chen. Y; *J. Phys. Chem. C.* Vol. 112, p. 17554, 2008
391. Yang X.Y., Wang Y.S., Huang X., Ma Y., Huang Y. and Yang R. ; *J. Mater. Chem.*, 2011 DOI: 10.1039/C0JM02494E,
392. Ya-Ping Sun, Zhou B., Lin Y., Wang W., Fernando K.A.S., Pathak P., Mohammed J. M., Harruff B.A., Wang X., Wang H., Luo P.G., Yang H., Kose M.E., Chen B, Veca L.M. and Su-Yuan Xie,*; J. Am. Chem. Soc.*, Vol. 28, No. 24, p. 7756, 2006
393. Yazyev O. V. and Helm L.; *Phys. Rev. B*, Vol. 75, p. 125408, 2007
394. Yazyev O.V.; *Phys. Rev. Lett.*, Vol. 101, p. 037203, 2008
395. Yazyev O.V. and Katsnelson M.I.; Phys. *Rev. Lett.*, Vol. 100, p. 047209, 2008

396. Yazyev O.V. and S. G. Louie; *Nature Materials,* Vol. 9, p. 806, 2010
397. Yazyev O.V. and Louie S.G.; *Phys. Rev. B,* Vol. 81, p. 195420, 2010
398. Yerushalmi R., Jacobson Z. A., Ho J. C., Fan Z. and Javey A.; *Appl. Phys. Lett.*, Vol.91, p. 203104, 2007
399. Yong Q.D., Pang H., Shuyan R., Chen C., Chi Y. and Ting Y.; *Carbon,* Vol. 64, p. 245, 2013
400. Yoo E., Kim J., Hosono E., Zhou H., Kudo T. and Honma I.; *Nano Lett.*, Vol.8, p. 2277, 2008
401. Youn S-W., Ogiwara M., Goto H., Takahashi M. and Maeda R.; *J Mater Process Technol.,* Vol. 202, p. 76, 2008
402. Zang H. and Liang R.; *The Spectrum*, Vol. 16, p.16, 2003
403. Zeng Q.O., Cheng J.S., Tang L.H.; *Adv. Funct. Mater.*, Vol. 20, p. 3366, 2010
404. Zhang, Y.; Tan, Y. W.; Stormer, H. L. and Kim, P.; *Nature,* Vol. 438, No. 7065, p. 201, 2005.
405. Zhang L. M., Xia J.G., Zhao Q.H., Liu L.W. and Zhang Z.J.; *Small,* Vol. 6, p. 537, 2010
406. Zhang L., Yu J., Yang M., Qin Xie, Peng H. and Z.; *Nature Communications* 4, Article No.1443, 2013, doi:10.1038/ncomms2464
407. Zhang Y.B., Ali S.F. and Dervishi E.; *ACS Nano.*, Vol. 4, p. 3181, 2010
408. Zhang Z., Hao J., Zhang J., Zhang B. and Tang J. ; *RSC Adv.*, Vol. 2, p. 8599, 2012,
409. Zhao Q.L., Zhang Z.L., Huang B.H., Peng J., Zhang M. and Pang D.W.; *Chem Commun*, Vol. 44, p. 5116, 2008,
410. Zhao L., Rim K. T., Zhou H., He R., Heinz T. F., Pinczuk A., Flynn G. W. and Pasupathy A. N.; *arXiv*, Vol. 1008, p. 3542, 2010
411. Zhao H. X., Liu L.Q., De Liu Z., Wang Y., Zhao X.J. and Huang C. Z.; *Chem. Commun.* Vol. 47, p. 2604, 2011
412. Zheng L., Ma L., Shi G., Zhou W., Yongji G., Lei S., Yang X., Zhang J.,Yu J., Hackenberg K. P., Babakhani A., Juan-Carlos Idrobo, Vajtai R., Lou J. and Ajayan P.M.; *Nature Nanotechnol* ; DOI: 10.1038/NNANO.2012.256
413. Zheng L.Y., Chi Y.W., Dong Y.Q.; J.Am.Chem. *Soc.*, Vol. 131, p. 4564, 2009
414. Zheng Y., Ni G.-X., Toh C.-T., Zeng M.-G., Chen S.-T., Yao K. and Ozyilmaz B.; *Appl. Phys. Lett.*, Vol. 94, p. 163505, 2009
415. Zheng X.H., Huang L.F., Wang X.L., Lan J. and Zeng Z., *Computational Materials Science*, Vol. 62, p. 93, 2012
416. Zhihua X., Jiaguo Yu and Liu G.; *Sensors and Actuators B,* Vol. 181, p. 209, 2013
417. Zhou J., Booker C., Li R., Zhou X., Sham T. K., Sun X. and Ding Z.; *J. Am. Chem. Soc.*, Vol. 129, No. 4, p. 744, 2007.
418. Zhou M., Zhai Y.M. and Dong S.J.; *Anal. Chem.* Vol. 81, p. 5603, 2009
419. Zhou Y., Q., L.A.L., Zhong Y. and Loh K.P.; *Chem. Mater*, Vol. 21, No.13, p. 2950, 2009
420. Zhou J., Booker C., Li R., Sun X., Sham T. K. and Ding Z.; *Chem. Phys. Lett.*, Vol. 493, p. 296, 2010

421. Zhou L., Lin Y., Huang Z., Ren J. and Qu X.; *Chem. Commun.*, Vol. 48, p. 1147, 2012
422. Zhou S. and Bongiorno Angelo; *Scientific Reports*, Vol. 3, p. 2484, 2013
423. Zhu H., Wang X., Li Y., Wang Z., Yang F., Yang X. *Chem Commun*, Vol. 45, p. 5118, 2009 Zhou J., Sheng Z., Han H., Zou M. and Li C.; *Mater. Lett.* Vol.66, p. 222, 2012
424. http://graphenea.com/blogs/graphene-news/8017795-graphene-mass-production-roll-to-roll#ixzz2fdEfsQrv
425. http://www.nobelprize.org/nobel-prizes/physics/laureates/2010/
426. http://Graphenea.com/blogs/Graphene-news/8017795-Graphene-mass-production-roll-to-roll#ixzz2fdEfsQrv
427. http://en.wikipedia.org/wiki/Graphane

Index

Also of Interest

Check out these published volumes in the Advanced Materials Series

Advanced Bioelectronic Materials
Edited by Ashutosh Tiwari, Hirak K. Patra and Anthony P.F. Turner
Forthcoming 2015. ISBN 978-1-118-99830-4

Graphene
An Introduction to the Fundamentals and Industrial Applications
By Madhuri Sharon and Maheswar Sharon
Published 2015. ISBN 978-1-118-84256-0

Advanced Theranostic Materials
Edited by Ashutosh Tiwari, Hirak K. Patra and Jeong-Woo Choi
Published 2015. ISBN: 978-1-118-99829-8

Advanced Functional Materials
Edited by Ashutosh Tiwari and Lokman Uzun
Published 2015. ISBN 978-1-118-99827-4

Advanced Catalytic Materials
Edited by Ashutosh Tiwari and Salam Titinchi
Published 2015. ISBN 978-1-118-99828-1

Graphene Materials
Fundamentals and Emerging Applications
Edited by Ashutosh Tiwari and Mikael Syväjärvi
Published 2015. ISBN 978-1-118-99837-3

DNA Engineered Noble Metal Nanoparticles
Fundamentals and State-of-the-art-of Nanobiotechnology
By Ignác Capek
Published 2015. ISBN 978-1-118-07214-1

Advanced Electrical and Electronics Materials
Process and Applications
By K.M. Gupta and Nishu Gupta
Published 2015. ISBN: 978-1-118-99835-9

Advanced Materials for Agriculture, Food and Environmental Safety
Edited by Ashutosh Tiwari and Mikael Syväjärvi
Published 2014. ISBN: 978-1-118-77343-7

Advanced Biomaterials and Biodevices
Edited by Ashutosh Tiwari and Anis N. Nordin
Published 2014. ISBN 978-1-118-77363-5

Biosensors Nanotechnology
Edited by Ashutosh Tiwari and Anthony P. F. Turner
Published 2014. ISBN 978-1-118-77351-2

Advanced Sensor and Detection Materials
Edited by Ashutosh Tiwari and Mustafa M. Demir
Published 2014. ISBN 978-1-118-77348-2

Advanced Healthcare Materials
Edited by Ashutosh Tiwari
Published 2014. ISBN 978-1-118-77359-8

Advanced Energy Materials
Edited by Ashutosh Tiwari and Sergiy Valyukh
Published 2014. ISBN 978-1-118-68629-4

Advanced Carbon Materials and Technology
Edited by Ashutosh Tiwari and S.K. Shukla
Published 2014. ISBN 978-1-118-68623-2

Responsive Materials and Methods
State-of-the-Art Stimuli-Responsive Materials and Their Applications
Edited by Ashutosh Tiwari and Hisatoshi Kobayashi
Published 2013. ISBN 978-1-118-68622-5